The Internet Guide For Dentistry

The Internet Guide
For Dentistry

by

Paul Downes

General Dental Practitioner and IT & CAL Tutor,
Thames Postgraduate Medical and Dental Education

1999

Published by the British Dental Association
64 Wimpole Street, London W1M 8AL

ISBN 0 904588 60 2

Printed and bound by
Dennis Barber Graphics and Print, Lowestoft, Suffolk

Foreword

Despite the fact that the 'electronic revolution' has been with us for some years, many people are still either mystified about, or terrified at the prospect of having to use, computers, e-mail, internet and associated applications. For those involved with dentistry this book provides a wonderful introduction to the apparent mysteries and the undeniable benefits of connecting to the internet. Indeed, it also provides a step-by-step explanation about how to connect.

Collected together from a series of articles in the *British Dental Journal*, the book is an essential guide for dentists, members of the dental team and dental students new to the internet, as well as providing a valuable reference source for those who already have some 'literacy' in matters of electronic communication. Each chapter is clearly written and easy to understand with the added benefit of having text of a more technical nature marked separately. This helps readers at different levels of understanding to navigate their way through the book and provides an excellent way of gradually increasing knowledge by revising chapters with steadily increasing degrees of comprehension.

The liberal illustrations of screens ensure that the reader is familiar with what to expect when transferring to the keyboard and monitor and complement the book's comprehensive overview of the subject. This book is set to become an essential item to be found by the keyboard of everyone connected with dentistry and to the internet, for both dental and non-dental applications.

Roger Farbey
Information Centre Manager
British Dental Association
London, January 1999

Preface

I started using the internet in early 1995, and initially found it quite a struggle to grasp what it was all about. At that time there was a shortage of books for the novice internet user, and those that were available were squarely aimed at the USA market. This book is written with UK dentists in mind, although much of the content is relevant to dental students and all members of the dental team, wherever in the world they work. The main theme is how dentists can make good use of this powerful technology. Most of the information is directed at the beginner, although there should be sufficient detail to satisfy the curiosity of the more experienced user.

This book is not designed as a comprehensive reference; think of it more as a starting point for a journey into the ever-changing world of 'cyberspace'. Some of the chapters are designed as practical lessons and it is intended that these should be read alongside using the relevant software program.

Considering how large and fluid the internet is, there are inevitably going to be some omissions or over-simplifications. My main concern was to make the subject clear and understandable. I would welcome any comments readers may wish to make.

I have included a short bibliography for those who wish to pursue this topic further.

Paul Downes
Whitstable, January 1999

Bibliography
Gibbs S, Sullivan Fowler M, Rowe N. *Medical Surfari*. Mosby, 1996.
Kiley R. *Medical information on the Internet*. Churchill Livingstone, 1996.
Schleyer T, Spallek H, Spallek G. *The global village of dentistry*. Quintessence Publishing Co, Inc, 1998.
Habraken J. *The big basics book of the Internet*. 2nd edn. Indianapolis, USA. Que Corporation, 1998.
Russell C. *Internet UK in easy steps*. 3rd edn. Warwickshire, UK. Computer Steps, 1998.

Acknowledgements

I am particularly grateful to David Speechley who scrutinised the contents of the initial *British Dental Journal* articles and gave me valuable constructive criticism.

Microsoft®, Windows®, Internet Explorer® are registered trademarks of Microsoft Corporation. Netscape®, Netscape Navigator® and Netscape Communicator® are all registered trademarks of Netscape Communications Corporation. Eudora Pro® and Eudora Lite® are registered trademarks of QUALCOMM Incorporated. All other trademarks are acknowledged as belonging to their respective companies.

Contents

1 An introduction to the internet

In September 1995, the leader article in the *British Dental Journal*[1] spoke about how the 'electronic revolution' would force education into rethinking its attitudes. New concepts such as the individual exploration for information and the active involvement in our own learning were discussed. The internet is now a major part of this electronic revolution and offers an ideal opportunity for the 'learning dentist' to search for information and communicate with other like-minded people. A number of books have been written about the internet for medical health professionals,[2–5] but little has been published specifically for UK dentists.[6]

Aim of this series

The aim of this series is to introduce the dental practitioner to the internet using as little technical jargon as possible. It explains how to connect to the internet and gives some examples of what you can do once you are connected or 'on-line'. There are plenty of practical tips on how to use the internet efficiently and how dentists can find information relevant to their needs. The last part of the series looks at current developments of this exciting technology, especially in relationship to dentistry.

The purpose of the series is to show how the internet is becoming an indispensable tool for the busy dental practitioner and to encourage colleagues to 'get on-line'. For those readers who are already connected to the internet or who want to delve into the subject more deeply, the areas of text with a vertical blue rule on the left-hand side are of a more technical nature. Avoid these if you are a technophobe!

What is the internet ?

The television, newspapers, magazines and radio have been bombarding us with news about the internet over the past couple of years. Although the topic has been subjected to a lot of media hype, it is true to say that the internet is here, and it is here to stay, but what exactly is it?

If you wanted to explain television to someone who had never seen it before, you would not start talking about cathode ray tubes, signal frequencies or transmitters. The same is true for the internet, so I shall start off with describing what it can do. The following list, which is by no means exhaustive, will give you a taste of what is possible once your home computer is connected to the internet by your telephone line:

- Send letters to friends in Australia, America and Asia for less than the cost of a second class stamp. They should be able to receive their letters within a couple of minutes.
- Enclose a picture, a database, or a financial spreadsheet with your letter (or any other type of computer file for that matter).
- Read today's newspapers free, (for example, *The Times*,[7] *Telegraph*,[8] *New York Times*,[9] and scores of others).
- In the *Electronic Telegraph*, all of their published material since November 1994 has been electronically archived. Search for every article that contains the word 'dentist', and you will find over 180 articles.
- Discuss the latest ideas about education, football, or politics with people from every corner of the world.
- Obtain free advice on almost any topic you could possibly think of.
- Track down the lyrics to your favourite West End show.[10]
- Have the opportunity to try out literally thousands of different software programs, and only pay for them if you find them useful.[11]
- Display an up-to-the-minute balance and statement from your home bank account.[12] Pay bills, view standing orders and direct debits and view transactions with a search and sort facility. You can also transfer your financial data into a financial management package such as *Microsoft Money 98*, plan your finances and print out bank details.
- Find out how much your shares are worth today and automatically update your financial accounting package (*Quicken 98*)[13].
- Give your children an excellent means of finding information for their homework, school project or hobbies. The BBC Education web site has an online service for teachers and parents, as well as details of school and college performance tables.[14]
- Send a record request to your favourite radio station.
- Buy some flowers from Interflora,[15] wine from Wine Cellars UK[16] or a CD from the 50,000 available at the Internet Music and Video Shop.[17] Place your order 24 hours a day.
- Find out what special events are taking place in your area next weekend.[18]

This is the first of a multipart series on the internet aimed at dental practitioners. In this part I will try to answer the questions 'What is the Internet?' and 'How does it work?', the historical background of the internet will be discussed, and reasons will be given as to why dentists should consider connecting to the internet.

Key:

| Text that is of a more technical nature

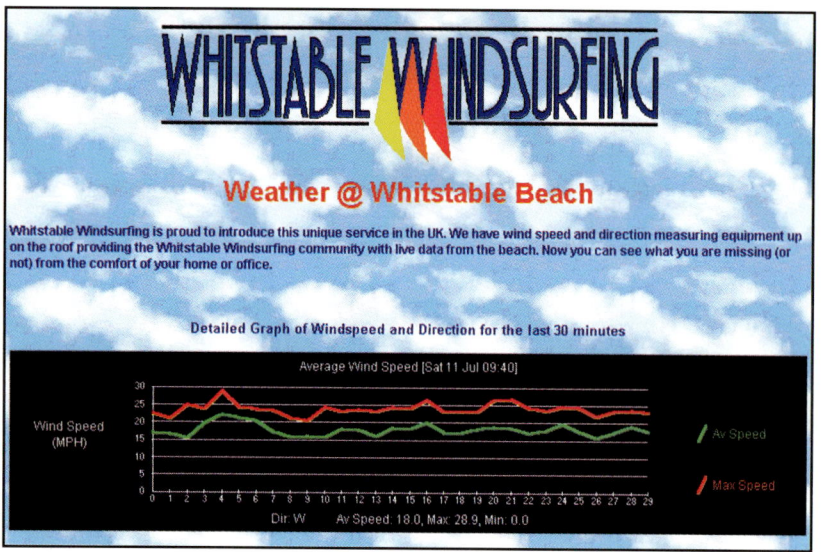

Fig. 1 Screen shot from the Whistable windsurfing website

How does information travel on the internet?

The computers which make up the internet are connected to one another in various ways: by satellite links, by optical fibres, by the integrated services digital network (ISDN) and by the humble telephone line, with its twisted pieces of copper wire.

In order for the many different types of computers to exchange information, a set of protocols called transmission control protocol/internet protocol (TCP/IP) have been established. TCP/IP breaks the data into small packets of information that can travel independently of each other through the networks of the internet. The packets of data are effectively placed inside a secure 'envelope' and labelled, in much the same way as you label packing cases when you are moving house.

The networks that comprise the internet are connected by computers known as 'routers', which decide how to transmit the data most efficiently across different parts of the network. The packets frequently arrive out of sequence because they travel through different pathways. The labels include information that helps reconstruct the complete message or file. If a packet of data is lost en route, the receiving computer requests another copy of the packet until it arrives intact.

The internet is responsible for what can only be described as a 'communication revolution'. Instead of thinking simply in terms of computer networks, think 'people networks'. Millions of people from all walks of life communicating not only on a one-to-one level, but also organised into self-selected groups for the sharing of information and ideas. These people make up the internet community.

What makes the internet so valuable is that it is an open network as opposed to a closed network. In other words, companies, governments, universities, societies, organisations, and individuals produce information which is, by and large, freely available to anyone connected to the internet.

The information is stored on the thousands of computers which make up the internet and consists of text, databases, interactive tables, sound, images and video. This global sharing of facts, figures and ideas is what makes up the often quoted 'cyberspace'.

The other exciting thing about the internet is that no one person or group of people is in charge. It is constantly evolving as people think of more novel ways of using the internet. There is no overall plan or blueprint for its development. The pace at which it is expanding and developing is breathtaking. A month is a long time in the life of the internet and a year can see enormous changes to almost every aspect of its structure.

- For those people interested in water sports, receive live data about wind speed and direction from a weather station located at a windsurfing shop on the south coast of England (fig. 1).[19]
- Plan this year's walking holiday in the Lake District[20] (fig. 2) or next year's skiing trip to the Alps.[21]

You can do all this and more from the comfort of your armchair for around £10 a month plus the cost of a local telephone call. If that sounds interesting, then maybe this series of articles is for you.

How does the internet work?

The internet is often described as the world's largest collection of computer networks. It is an affiliation of tens of thousands of private, commercial, academic and government-supported computer networks from nearly every country in the world. Almost any computer can connect to it, from powerful mainframes to simple home microcomputers such as the *IBM-PC* and *Apple*. Special software allows all the computers to speak the same language.

Fig. 2 Screen shot from the Lake District Walks website

Internet applications

The internet is made up of various applications, (for example, e-mail and the world-wide web), and these can be thought of as the various tools by which you access information on the internet. These applications all have their own software programs that you install on your home computer. The programs allow you to send and retrieve information to and from the internet.

The most popular internet applications, and what they do, will be discussed in more detail later in the series. They are as follows:

• E-mail — used in writing electronic letters.
• World-wide web (WWW) — used for browsing the pages of information held on the internet.
• Usenet (newsgroups) — used to participate in worldwide discussions.
• File transfer protocol (FTP) — used to retrieve free or inexpensive software from the internet.

How did the internet come into being?

It is true to say that the internet evolved rather than it being invented. It all started in 1969 with the ARPANET (the Advanced Research Projects Agency Network). This was a US government funded project which experimented with linking computers from different parts of the country (networking). The aim was to provide researchers and scientists with access to all of the machines, programs and data on the network.

To enable the rapid flow of information around the network, the data was broken down into small, manageable sized 'packets'. The computers had the ability to re-route these packets of data in the event of the message not being able to be transmitted by its chosen route, for example, if the connection was busy or broken.

By the mid-1970s, selected educational and research institutions were invited to join the network. It became hugely popular with universities for sharing information and ideas. Corporate businesses then became involved as they could see the potential for improving their own communications. People soon came to realise that the real power of the internet was not that the computers were able to talk to one another, but that people could talk to one another.

The meteoric explosion in popularity of the internet is a more recent phenomenon and is because of the graphical nature of the world-wide web. In December 1993, a piece of internet software called *Mosaic* was developed; the software was free and within six months, more than two million people had downloaded it onto their computers to use. The reason for this was that it allowed you to browse the information on the internet with your mouse and point-and-click at links which would take you to various sites half way around the world. Web sites were soon able to display not just text, but also

graphics and photographs. Later it was possible to play animation, videos and sound. This made the WWW the most visual part of the internet and it really caught the attention and imagination of the media and public alike.

The most recent milestone in the history of the internet involves one of the richest men in the world. In 1994, Bill Gates, who is the chief executive officer of Microsoft, decided to channel a considerable part of the company's money, skills and resources into the development of the internet. Microsoft now has a significant influence on the way in which the software and hardware for this new technology will advance.

How big is the internet ?

The ARPANET project started in 1969 with just four supercomputers. Estimates now put the number of hosts, or individual computers connected to the internet, at more than 29 million; most of these have been added in the past 7 years. As of July 1998, Germany had the largest number of hosts in Europe, with over 1,287,000 computers connected to the internet. The UK came a close second with over 1,232,000 hosts,

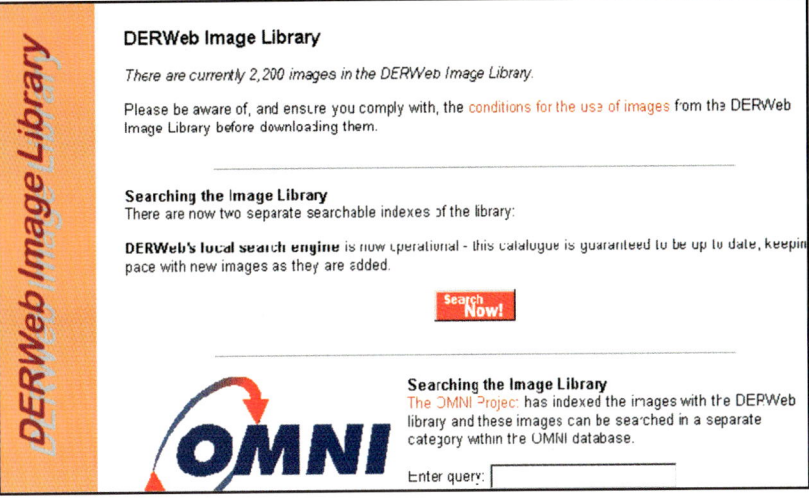

Fig. 3 This screen shot shows the introductory page to the *DERWeb* Image Library

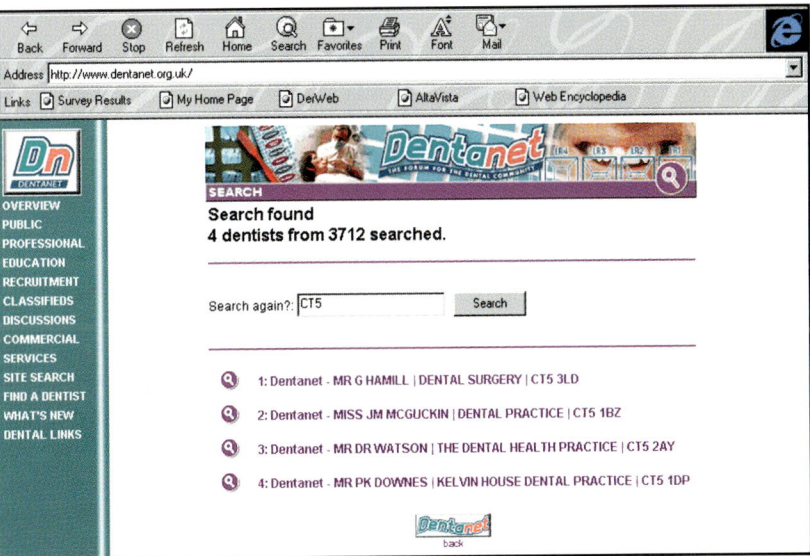

Fig. 4 This shows the result of searching for my own postcode in the 'Find a Dentist' section on the Dentanet website

the internet guide for dentistry

1 Grace M. The electronic revolution. *Br Dent J* 1995; **179**: 153.
2 Lee N, Millman A. *ABC of medical computing*. BMJ Publishing Group, 1996.
3 Pallen M. *Guide to the internet*. BMJ Publishing Group 1996.
4 Kiley R. *Medical information on the internet*. Churchill Livingstone, 1996.
5 Gibbs S, Sullivan-Fowler M, Rowe N. *Medical Surfarai*. Mosby, 1996.
6 Williams M. Step-by-step to the internet. *The Dentist*. May and June 1997.
7 *The Times*: http://www.the-times.co.uk/
8 *The Electronic Telegraph*: http://www.telegraph.co.uk/
9 *New York Times*: http://www.nytimes.com/
10 Tower Lyrics Archive to West End Shows: http://www.ccs.neu.edu/ home/tower/lyrics.html
11 WinFiles.com Windows Shareware Collection: http://www.winfiles.com/apps/
12 Royal Bank of Scotland: http://www.royalbanksscot.co.uk/
13 Quicken: http://www.intuit.co.uk/ investments/download.htm
14 BBC Education: http://www.bbc. co.uk/education/
15 Interflora: http://www.barclaysquare.co.uk/
16 Wine Cellars UK: http://www.winecellar.co.uk/
17 Internet Music and Video Shop: http://www.musicshop.co.uk/
18 British Tourist Authority: http://www.visitbritain.com/
19 Whitstable Windsurfing: http:// www3.mistral.co.uk/whitwind/
20 Lake District Walks: http:// www.netlink.co.uk/users/ldnet/
21 The Ultimate Ski Links Page: http://www.geocities.com/Yosemite /9818/
22 European Hostcount: http://www.ripe.net/statistics/hostc ount.html
23 Derweb: http://www.derweb.ac.uk/
24 *Journal of Clinical Pediatric Dentistry*: http://www.pediatric dentistry.com/journal.html
25 *Dental Bytes*: http://www. sybor.com/dentalbytes/
26 Newsgroup related to dentistry: sci.med.dentistry
27 *Dentanet*: http://www.dentanet.org.uk/
28 Classic Dental Center: http://www.classicdental.com/
29 The Interactive Investor: http://www.iii.co.uk/
30 PubMed, National Library of Medicine: http://www4.ncbi. nlm.nih.gov/PubMed/
31 OMNI: http://www.omni.ac.uk/

Most of these references refer to a uniform resource locator (URL). The URL can be thought of as the 'address' of a page of information on the WWW. URLs will be discussed more fully later in the series.

and in third place was Holland with 488,000 hosts.[22] It is estimated that more than 50 million people have access to the internet, and that the majority of these people live in the USA.

When Sir Paul McCartney appeared in a live global internet link-up, the former Beatle was bombarded with 2.6 million questions from fans. The popularity of this form of communication cannot be underestimated. The internet is larger than the sum of its parts and is incessantly evolving into new and surprising forms.

Why should dentists consider connecting to the internet?

As well as the many recreational uses of the internet, there are also many ways in which dentists can specifically benefit. The following list will help to show some of these dental and business applications:

- There is a vast amount of dental information, (both clinical and non-clinical), available on the internet. DERWeb[23] is just one of many UK sites that contain a huge quantity of useful dental material including over 2200 dental pictures, see fig. 3. I have put this service to use in supplementing the stock of digital images which I use for patient education in the practice.
- Many traditional dental journals are now published 'on-line'.[24] New electronic dental journals, which only appear in their electronic form, have also developed.[25]
- Information is often published on the internet long before it appears in print.
- There are dental-related discussion groups[26] which you can subscribe to free of charge. If you want to find out how other dentists around the world operate their own businesses, you will find plenty of people willing to converse.
- E-mail is a fast, efficient and cost-effective form of communication, especially overseas.
- The internet is the most efficient way of obtaining free updates for some of the software that you may have already purchased. In one case I found that it was the only way in which I could receive a software update, since it was company policy to no longer send out floppy disks through the post.
- Dentists are able to advertise and promote their practices to the general public. *Dentanet* is in the process of setting up a list of all UK dentists from which potential new patients can search by their own postcode, (see fig. 4).[27] For a small fee, *Dentanet* or *DERWeb* will create a 'web page' about your practice, which can include such things as opening hours, photographs of the practice staff or even a map.
- It is possible for dental practices to have quite sophisticated web pages. There is a practice web site where existing patients can make appointments, fill in an anonymous 'satisfaction questionnaire' about the practice and search more than two hundred fully illustrated pages of dental terms and procedures.[28]
- There are an increasing number of excellent UK financial sites which give free advice on money matters, such as taxation, investments and borrowing.[29]
- There are places on the internet which provide free access to *Medline*,[30] the largest biomedical reference library in the world, containing more than 9 million abstracts from 3800 medical journals.
- The quality of the information on the internet is improving all the time and the number of UK-related sites has increased dramatically. There are now many specialised healthcare sites which index the most useful medical and dental resources,[31] making it much easier to find the information that you are looking for. You also have the reassurance of knowing that the content of these resources have already been scrutinised.

There has never been a better time to start using the internet. Compared with just a few years ago, it is now very easy to set up a connection. Increased competition has brought down the cost of hardware and subscriptions, while increased modem speeds have made the task of 'surfing' the web less of a wait. At the moment most of the content on the internet is free, but it will not be long before some of the really useful commercial sites will start charging for access to them. It is quite possible that the internet has now reached its 'critical mass', where there are enough people using the technology to make it a worthwhile communication tool for everyone; now is the time to become internet-literate.

2 Connecting to the internet

To obtain a connection to the internet a number of requirements need to be fulfilled. This list is actually quite short: computer, modem, telephone line, internet account with an internet access provider, and internet software. We shall now consider each of these items in more detail.

Computer

This is the biggest cost involved in connecting to the internet, but of course the same computer can be used for numerous other tasks related to both work and recreation. Many dental practitioners who already use the internet do so from home. It is often the case that the specification of their home computer is higher than the machine(s) used in the practice.

During this series, I will limit the focus of discussion to the IBM-compatible PC, since this is the most popular form of personal computer. However, as mentioned earlier, almost any type of personal computer, (such as the Apple Macintosh, Acorn Archimedes or Amiga), can be connected to the internet.

Table 1 shows a typical mid-level specification for a multimedia PC which will enable the user to take full advantage of most of the internet applications currently available. (Multimedia means that the computer can play sound and moving pictures.)

If you are considering buying a new computer, then the good news is that prices continue to tumble as computing power increases. A mid-level PC currently costs (prices quoted July 1998) in the range of £800–£1200. Expect to pay around £600–£700 for an entry-level multimedia PC and more than £1500 for a state-of-the-art home multimedia PC, (which will only remain state-of-the-art for a matter of a few months, such is the speed of technological development).

Modem

To send and receive data through the internet you require a modem. Anyone who transmits National Health Service claims to the Dental Practice Board by Electronic Data Interchange (EDI) is already using a modem.

A modem can either be built into the computer as an internal card or it can sit on your desktop as an external box. The advantages of an internal modem are that it takes up no space on your desk and does not require a separate power supply. On the other hand, an external modem is portable and can therefore be used with more than one computer. It is often more economical in the long term to purchase one of the higher speed modems, since this will reduce the time spent downloading data while you are on-line. This in turn will reduce your telephone bills.

Most modern modems include a faxing facility; this means that your computer will be able to send faxes without having to print them onto paper first. Provided your computer is switched on, it will also be able to receive incoming faxes. Many new computers are being sold as 'internet ready', which means that the computer comes with an internal modem as well as all the software you need to subscribe and access the internet.

More about modems

The word 'modem' stands for **modulator-dem**odulator. Computers communicate internally by digital signals (a series of 1's and 0's). The telephone network is designed to carry analogue, or speech-like frequencies. A modem translates the digital data from a computer into analogue data, so that it can be sent down the telephone line. At the other end of the telephone line, the receiving computer's modem then converts the analogue data back into digital data. Most modern modems have automatic error correcting capabilities to deal with poor quality telephone lines.

The most popular speed modems at the moment are 33.6 Kbps and 56 Kbps, (thousand bits per second). Some older modems used in dental practices may be considerably slower than this and are not suitable for connecting to the internet.

The cost of modems has dropped dramatically in the past few years. I would recommend any prospective purchaser to buy a 56 Kbps speed modem since these can now be bought for less than £100. Make sure that it is fully compatible to the new V90 specification. When purchasing a modem, pick a reliable brand name such as Motorola, US Robotics, Hayes, Pace, Psion, or Novatech. Some of the latest, more expensive modems are upgradable by using a feature called flash ROM.

Telephone line

A standard telephone line is all that you require as a physical means of connecting to the inter-

This second part of the internet series explains what is required to obtain a connection to the internet and the likely costs involved. Advice will be given on how to pick a suitable internet access provider. The process of getting on-line for the first time will also be discussed.

Key:

Text that is of a more technical nature

password
Word(s) typed in at the keyboard

web browser
Keyword defined in the margin

Table I. A typical mid-level specification for an IBM-compatible PC multimedia system

Specification
Windows 98 operating system
Pentium 200 MMX processor
16–32MB RAM
15" screen
2GB+ hard disk
12+ speed CD-ROM drive
Soundblaster compatible sound card and speakers
Modem (56 Kbps)

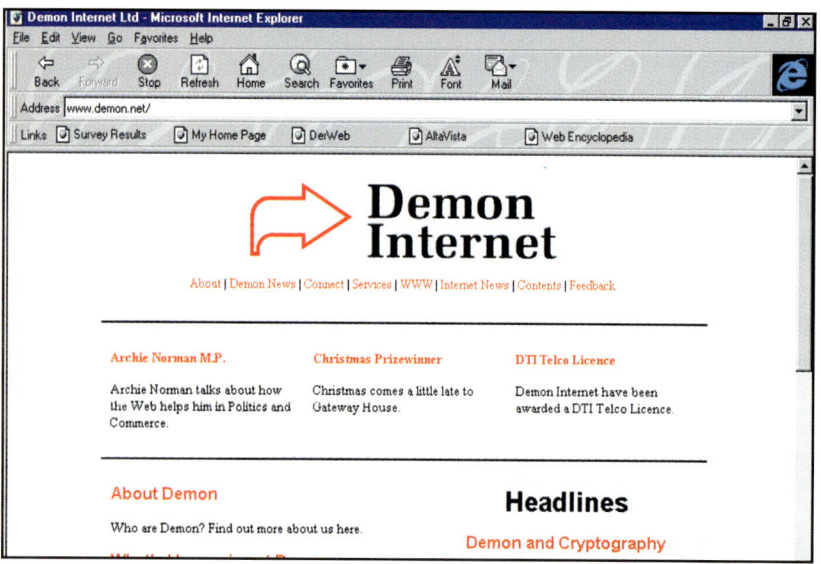

Fig. 1 Screen shot of the home page of the internet service provider, Demon Internet, as viewed through a web browser

Fig. 2 Screen shot showing the sort of special content provided by an on-line service (CompuServe), for their members

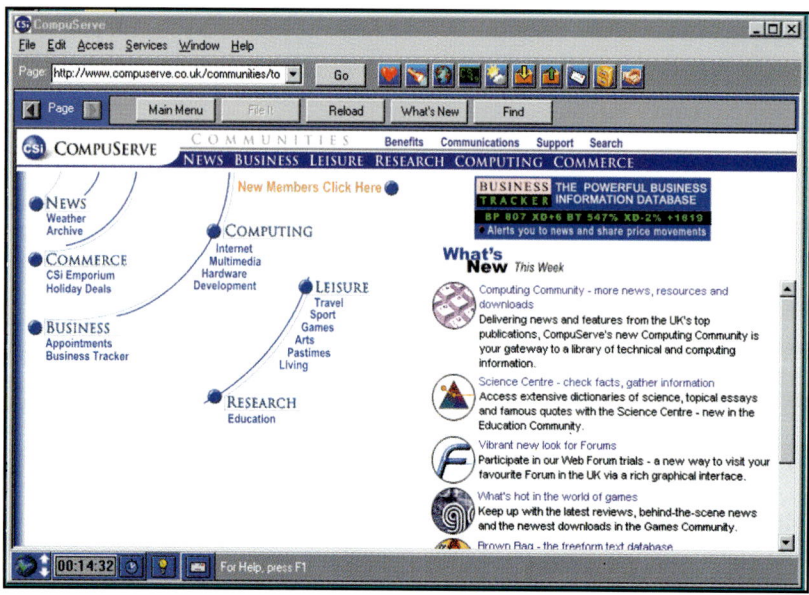

net. Having a good quality line improves the reliability of the connection. While you are on-line, no one will be able to dial in or out on that telephone line. Assuming that you will be using a local call rate, the cost of being on-line will depend on how many minutes you are using the phone and at what time of day you are calling. Most home users access the internet in the evenings and at weekends, resulting in charges of around 1p per minute.

ISDN

ISDN stands for Integrated Services Digital Network. It is a digital voice and data telephone network which looks set to replace the current analogue one. ISDN adapters are already starting to replace modems as a fast method of accessing the internet. ISDN is often used by big businesses for video conferencing, transferring large files and remote network access. However the high cost of line rental means that it is still out of reach for most PC users and small businesses. ISDN supports transfer rates of 64 Kbps, but most telephone companies offer you two lines at once, called B channels. You can use one line for voice and the other for data, or you can use both lines for data which gives you a transfer speed of 128 Kbps.

Internet access providers

The most popular method for individuals to connect to the internet is by paying for an account with either an internet service provider (ISP) or through an on-line service. In effect, these companies sell you a way into the internet through their own high speed computers and connections.

Internet service provider

You can sometimes find an up-to-date list of ISPs in one of the national internet magazines.[1,2] There are currently over 100 ISPs operating in the UK and some of the smaller service providers just serve specific regions of the country. Two of the biggest nationwide ISPs are Demon and UUnet Pipex. See figure 1 for an example of what an ISP home page looks like when viewed through a web browser.

On-line service

On-line services offer internet access in addition to their own exclusive services, such as reference information, computer games, local 'What's On' listings and chat areas that are only accessible to members. A good example of this is the new *Which?* On-line service,[3] where members can access all of the independent consumer advice from the whole range of *Which?* magazines. On-line services often have parental controls to limit access to adult material. Other examples of on-line services in the UK are America On-line (AOL), CompuServe, Microsoft Network (MSN) and Virgin. It is common for computer magazines and newspapers to give away a free CD-ROM or diskette

Table 2 Details of some popular internet access providers. Quoted prices are inclusive of VAT and were correct on 20/07/98

	Limited rate of hours per month	Additional hours	Unlimited time on-line, per month	Set-up fee	Telephone number
On-line services					
AOL	£4.95 for 3 hours	£2.35	£16.95 (or £14.95 if paid in full for 1 year)		0800 279 1234
CompuServe: Standard	£6.50 for 5 hours	£1.95			0990 000 200
Flat rate pricing plan			£17.95		
MSN: Hourly pass	£4.95 for 3 hours	£1.95			0345 002 000
Monthly pass			£14.95 (or £12.50 if paid in full for 1 year)		
Virgin Net			£11.99		0500 558 800
Internet service providers					
Demon			£11.75	£14.68	0181 371 1234
UUNet Pipex Dial			£14.98	£13.51	0500 474 739

which contains software to subscribe to one of the on-line services. They normally allow you to subscribe for a free trial period, (but be prepared to quote your credit card details). See figure 2 for an example of the sort of special content created by an on-line service for their own members.

Most ISPs charge a set-up fee of about £14, and then a flat monthly fee of £12–£15. The on-line service companies either charge a flat monthly fee of about £12–£18, or else charge a smaller monthly fee which includes 3–5 hours on-line per month and then charge extra for each additional hour, (see Table 2 for details). Many providers supply a guide or tutorial for using the internet, either on a CD-ROM or on-line.

The additional elements provided by on-line services invariably means that they are slightly more expensive than connecting to the internet through an ISP, but because of increased competition, the difference is not as great as it used to be. If you only envisage using the internet for a few hours each month then picking an on-line service may work out cheaper. It is true to say that most of the content and services found on on-line services has now been duplicated on the internet, only you have to go out and find it for yourself. If the internet were a bicycle, then an on-line service is the internet with training wheels. If you choose to open an account with an ISP then I would recommend joining one of the larger providers. There is no guarantee that the smaller operators will survive the increasing competition, and it could be a bind to later have to change your e-mail address.

Providers' options change almost monthly so it is worth staying in touch with developments to see which company is offering the best deal in terms of cost, ease of set up, software supplied, connection speed and technical support.[4] Make sure that your chosen provider offers local-call access to the internet since the cost of telephone calls can make up a significant part of the total cost of using the internet.

PoPs

When you connect to the internet your modem dials one of your access provider's access nodes. These are normally called Points of Presence or PoPs. To keep down the cost of your telephone bill, you should choose an access provider which has a PoP within a local telephone call from your computer. This does not mean that the access node needs to be situated in or around your town. The larger access providers have entered into agreements with telephone companies to treat PoP calls on 0345 or 0845 numbers as a local call, no matter where in the UK it is dialled from.

You may also want to take into account whether or not your access provider supplies you with any free space on their computers for

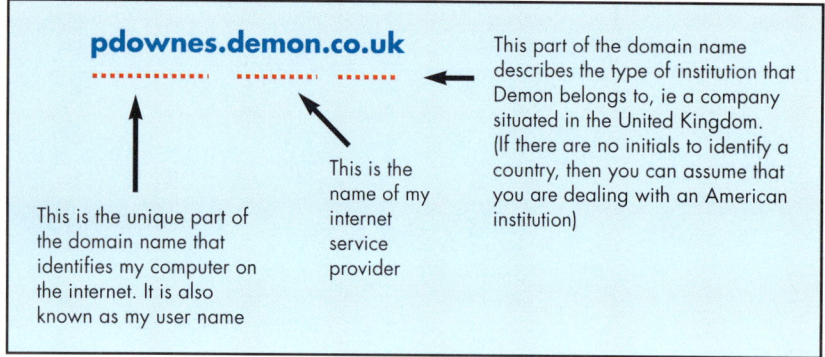

pdownes.demon.co.uk

This is the unique part of the domain name that identifies my computer on the internet. It is also known as my user name

This is the name of my internet service provider

This part of the domain name describes the type of institution that Demon belongs to, ie a company situated in the United Kingdom. (If there are no initials to identify a country, then you can assume that you are dealing with an American institution)

Fig 3. Domain name

Web browser = software program that enables you to view pages on the world-wide web

E-mail = 'electronic mail': a way of sending messages between computers

Fig. 4 An example of one of the screens that guide you through the process of signing up with an on-line service

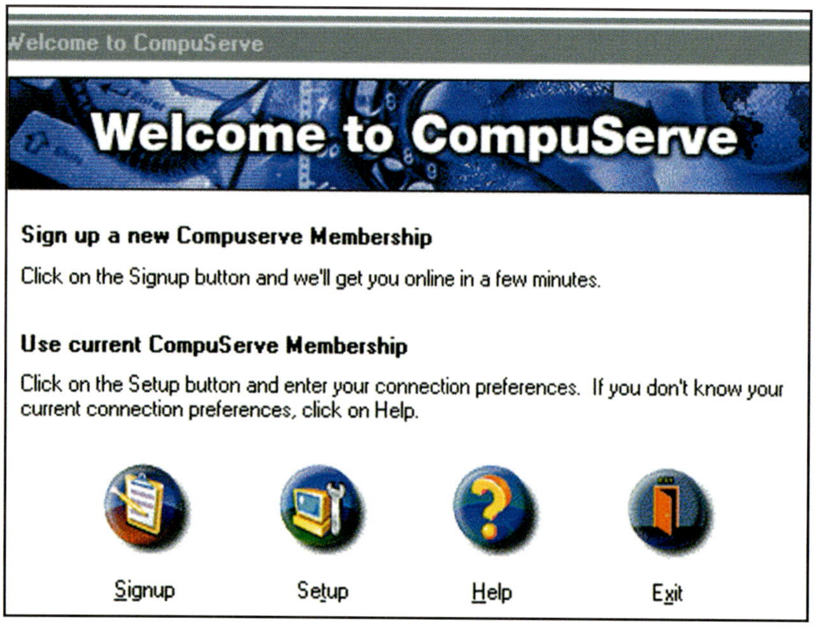

your own home web page. This is a place on the world-wide web (WWW) where you can advertise your existence: personal or business. You may not want to create a home web page straight away, but many people do so once they have become familiar with the internet. The amount of free web space supplied varies from 0.5–10 MB. (To give you an idea of what this means, 2 MB of space is enough to display 400 A4 pages of text; graphics take up appreciably more space.)

Internet software

Your chosen internet access provider will supply you with the necessary programs to get your computer on-line, plus some basic internet application software such as an e-mail program and a web browser. These programs are normally supplied on a floppy disk or CD-ROM as part of your membership. The installation programs from the major providers are very easy to use and it is normally a matter of answering some questions which appear on the screen and clicking the appropriate buttons.

How to 'get on-line'

Assuming that you have acquired the five items listed above, the next stage is to actually connect to the internet. Here is an overview of

how to go about it; the actual procedure will vary depending on which access provider you have chosen.

Step 1

Sort out an e-mail address and domain name for your computer.

With more than 30 million computers connected to the internet, it is obvious that you need a system to help locate each individual computer and user. The domain name is the name of the user's internet system or location. Some access providers will automatically assign you with a domain name and e-mail address; others allow you to pick one. For example, the domain name (fig. 3) of my computer is:

pdownes.demon.co.uk

Domain name system

The internet uses both a system of names (domain name) and numbers (IP address) to identify computers and individuals on the internet. The name and the numbers represent the same address. Computers use the number system to route information around the network, while people prefer to use names because they are easier to remember and mean more to them. You are supplied with your own unique domain name and IP address by your access provider, (depending on which access provider you are with, this number may change every time you log on).
The domain name may have one or more additional sub-domains which are used to group computers together. These sub-domains can tell you something about the person or institution you are dealing with. American sites can be recognised by **.com** = commercial, **.edu** = education, **.org** = non-profit making organisation and **.gov** = government. Educational sites in the United Kingdom can be recognised by **.ac.uk** where the **.ac** stands for academic.
Domain names create a single identity for a series of computers used by a company or an institution. So while there may be 38 servers at a given company, each with its own IP address, they all share a common domain name, such as **microsoft.com.**

Step 2

Contact your chosen access provider and ask them to set up an account. Here are two common ways of doing it:

• Ring the sales telephone number for your chosen access provider (see Table 2). You will be asked for your relevant personal details and, depending on the provider, your account should be set up within a couple of hours. You will be asked what sort of computer you are using, for example IBM-compatible or Apple Macintosh. The relevant

Welcome to CompuServe

Welcome to CompuServe

Sign up a new Compuserve Membership

Click on the Signup button and we'll get you online in a few minutes.

Use current CompuServe Membership

Click on the Setup button and enter your connection preferences. If you don't know your current connection preferences, click on Help.

Signup   Setup   Help   Exit

installation software disk or CD-ROM will be dispatched to you by post.

- If you use one of the cover-mounted disks from a computer magazine (or one boxed with your new modem) you run the installation software first. This will then automatically connect you to your provider through your modem and you will be prompted for your details by the installation software. (*Windows 95* includes a group of simple set-up buttons to enable you to join MSN (Microsoft's own on-line service), AOL or CompuServe.)

Step 3

Set up the internet software given to you by your internet provider by following the instructions as they appear on the screen; see figure 4 for an example. Make sure you have enough free space on your hard disk first and close down any unnecessary programs which may be running on your PC.

Step 4

The installation program will offer you a selection of telephone numbers for you to use when you go on-line. Select the provider's telephone number which will give you access to a local call rate. You will be asked if you have any special dialling properties; for example, if you need to dial 9 for an outside line.

Step 5

You will also be asked for a password which you later use whenever you go on-line.

Step 6

Most providers will bill your credit card on a monthly basis. Remember that many will give you a free trial period and there is normally no obligation to continue once your free period has expired; just remember to cancel the account.

If you experience any problems with setting up your internet connection, most providers have a technical support section who will be able to help you with the procedure. If you have access to a second telephone, (for example, a mobile phone), then they will even be able to talk you through the procedure as you go on-line.

Once your connection has been set up, there should be an icon on your desktop which you click whenever you want to go on-line. This will automatically dial and connect to your access provider.

Windows 95

If you are running *Windows 95* you need to make sure that the dial-up networking option has been installed. *Windows 95* comes with all the necessary communication protocol software to link your modem to the internet through a telephone line:

Table 3. Typical information about your internet account and connection details which you should record and keep in a safe place

Useful information	For example
Internet provider's name	Enterprise Internet Access Ltd
Helpline telephone number	0845 999999
Connection telephone number(s)	0845 123456
Connection type (SLIP or PPP)	PPP
Provider's domain name server address	158.234.2.19
Your account (host) name	jamestkirk
Your IP address (if supplied)	158.234.134.555
Your login password (KEEP THIS SAFE)	BeamMeUp
Your e-mail address	jamestkirk@enterprise.com
Mail server address	post.enterprise.com
POP3 address (if applicable)	pop3.enterprise.com
News server address	news.enterprise.com
WWW proxy server address and port (if applicable)	www.enterprise.com and 8080

TCP/IP, was covered in part one of this series. PPP, (point to point protocol), and its alternative SLIP, (serial line internet protocol), are used by different internet access providers. Ask your provider which protocol you should use. Both PPP and SLIP enable you to connect directly to the internet so that you can use programs on your own computer. ('Terminal' or 'dial-up' services only give you an indirect internet connection, which limit you to using programs on the service provider's computer.)

When you first open your account it is advisable that you keep a record of your account and connection details, (see Table 3). These details should be available from your internet provider and can later be used to configure additional internet software.

Universities, libraries and cybercafés

Another way of connecting to the internet is by borrowing or hiring someone else's internet connection.

Many dental schools now have computer clusters connected to the internet specifically for use by students and staff. It is also becoming increasingly common to find an internet connection at your local town library or postgraduate medical centre library. Ask your regional dental postgraduate administrator if there are any hands-on courses on using the internet.

The British Dental Association Library Multimedia Room has a PC connected to the internet for members to use. They charge £2.50 for half an hour on-line and novices are given a demonstration of how to get started on the internet.

Cybercafés are fun place to 'surf' the net and enjoy a cup of coffee at the same time. There are more than 50 cybercafés in the UK,[5] and you can find the address of your nearest one from looking in the listings which are sometimes found in the back of internet magazines.[1,2] They generally offer a degree of hands-on training and are an excellent way of trying before buying. A typical charge is £2.50 for half an hour on-line.

1. *Internet Magazine*, London: EMAP Business Communications.
2. *.net*, Bath: Future Publishing Ltd.
3. *Which?* On-line. Customer Services. Tel: 0645 830 240
4. UK and Irish Internet Company Directory: http://www.limitless.co.uk/inetuk/
5. List of cybercafés: http://www.ukdirectory.com/computer/cyb.htm

Notes:
There are many excellent books available which describe in detail about how to connect to the internet; two that I would particularly recommend are:
1. Habraken J. *The big basics book of the Internet.* 2nd ed. Indianapolis, USA. Que® Corporation, 1998.
2. Russell C. *Internet UK in easy steps.* 3rd ed. Warwickshire, UK. Computer Step, 1998.

3 E-mail: what is e-mail?

E-mail stands for electronic mail and it is simply a way of sending a message to another person over a computer network; in the case of the internet the message is sent to the other computer through the modem. It is the most widely used service on the internet, mainly because it is fast, cheap, informal and easy to use. There is no limit to the amount of messages that can be transmitted at any one time; numerous messages can be sent all over the world for the cost of a local phone call.

As well as sending plain text in your e-mail message, it is also possible to include pictures, (in colour, as well as in black and white), sound files, data such as a document, database or spreadsheet, or even a piece of software. Figure 1 shows the popular e-mail program, *Eudora Light*.

The advantages of e-mail over other forms of communication

Advantages of e-mail over the postal service (called 'snail mail' by users of the internet)
- Cheaper: especially for overseas communication. It costs the same to send a message to Tokyo as it does to Tottenham.
- Faster: especially for overseas communication.
- Easier to send a communication to multiple recipients, be it 20 or 2000 people.
- Quicker and less formal than writing a letter.
- No need to buy paper, envelopes or stamps.
- No need to visit the post office or post box.
- No need to copy data onto a floppy disk in order to send it to a recipient.
- Can send your communication 24 hours a day, 7 days a week.
- With e-mail, recipients of messages do not need to be near their computers to collect the messages. They can log onto their mailbox from anywhere in the world to pick up their messages.
- Can be used for group discussions, (see part 4 of the series).

Advantages of e-mail over a voice telephone call
- Cheaper; especially for overseas communication.
- More convenient, (often you cannot contact the person you want to speak to by telephone).
- You have a record of the communication.
- You can include a document, database or spreadsheet which the recipient can edit.
- You can communicate at a time which is convenient to you.
- Avoids the scenario of having to leave a message with someone else. The person invariably either does not receive the message or else returns your call when you are either out or busy.
- When communicating overseas, it is not necessary to take time zones into account.

Advantages of e-mail over paper fax
- More private and secure.
- Some faxes are very difficult to read because of the poor quality of their reproduction. Fax paper is difficult to store and deteriorates with time.
- Do not encounter paper jams, running out of paper or other fax errors.
- Do not waste paper, yet with e-mail you can print your important messages onto quality paper if required.
- Can send coloured pictures.
- Can attach a sound file to your message.
- Can include a document, database or spreadsheet which the recipient can edit.
- The fax machine needs to be turned on all the time to receive a fax, whereas you do not need to have your computer switched on for someone to send you an e-mail.

Disadvantages of e-mail
- Cannot send something physically by e-mail.
- Not everyone has an internet connection.
- It can be harder to find someone's e-mail address than their telephone number or postal address.
- Can sometimes be difficult to log onto the internet at peak times.
- Unlike the telephone call, you cannot have an instant 2-way communication.

Considering the advantages of e-mail over other forms of communication, it is therefore not surprising to find that in a recent survey on the use of the internet by dentists[1] 85% of the respondents considered e-mail to be one of their main uses of the internet, (compared to only 47% for the world-wide web).[1]

How does sending and receiving e-mail messages work?
You use an e-mail program, (which looks similar to a simple word processor), to send and

This third part of the series explains what e-mail is and why it is the most widely used service on the internet. The principles behind sending and receiving e-mail messages are described. Also the various uses of e-mail are discussed.

Key:

| Text that is of a more technical nature

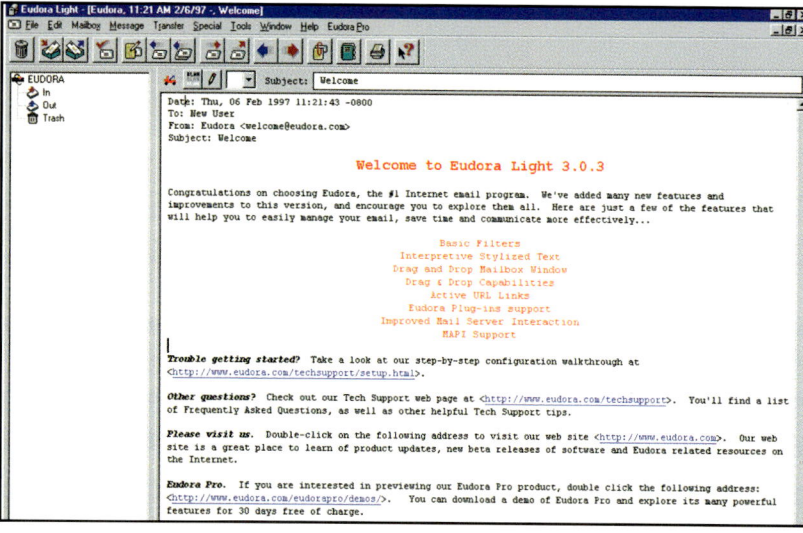

Fig. 1 The popular e-mail program *Eudora Light*

receive e-mail messages. You would normally compose a message on your e-mail program before you connect to the internet; by composing 'off-line' you avoid wasting money on unnecessary telephone charges. Once you have written your message or messages, they are stored in an 'out box', ready to be sent next time you go on-line.

When you decide to go on-line, most e-mail programs are configured to automatically send the messages waiting in your 'out box' and receive any messages which have been sent to you. As the incoming messages arrive, they are normally stored in an 'in box' for you to either read while you are still on-line or to read after you have disconnected from the internet.

The way this system works is that all your incoming e-mail is stored on a computer by your internet provider, (this computer is called a mail server). Your computer does not need to be switched on all of the time; your incoming messages will wait on the mail server until you go on-line to collect them. In the same way, the e-mail you send to other people first goes to your provider's mail server, which in turn sends it to your recipient's mail server. Here it is stored ready to be downloaded next time they go on-line.

SMTP and POP

All e-mail is sent using Simple Mail Transfer Protocol, (SMTP). However, there are two different methods for receiving e-mail; SMTP or Post Office Protocol 3 (POP3). POP3 is a protocol for receiving e-mail which is faster and more flexible than SMTP; it enables you to sort, delete or leave mail at the server end.

Not all access providers support POP3, and instead rely on SMTP for both sending and receiving e-mail. In order for you to use certain e-mail programs, (such as Eudora), check that your provider supports POP3.

Other uses of e-mail

E-mail also enables you to participate in group discussions, (mailing lists), and to receive regular updates on information, (mail newsletters):

Mailing lists

Mailing lists are subject specific discussion groups that are participated in and distributed by e-mail. There are more than 13 000 mailing lists you can subscribe to and they require no additional software other than your e-mail program. (Usenet newsgroups are similar to e-mail mailing lists in that they provide a means of joining in with group discussions; they will be discussed in Part 5 of this series).

Mailing lists work in one of two ways. Either a copy of every message sent to the list is forwarded to each member of the list, or else a digest of the messages is created by an administrator and sent out to the members on a regular basis, (normally daily).

Mail newsletters

E-mail is used for sending out information about a particular topic on a regular basis. This is a one-sided communication service. An example of this would be The Internet Bookshop,[2] who send me a regular e-mail newsletter about special offers as well as informing me whenever new books about dentistry are published.

Details on how to join or 'subscribe' to a mailing list will be discussed in the next part of this series.

E-mail mentality

One of the reasons why e-mail is so popular is its informality; it is quite acceptable to use a less formal style than used in writing letters. The content of your message may only be as short as one or two lines; this brevity looks quite out of place in a normal letter.

If you are replying to someone's earlier message then you simply click on a reply button to set up a correctly addressed reply. The date and your own e-mail address are also automatically added to any message you send. Quoting part or all of the body of their message is also very easy to do and can save a lot of typing on your part. The quoted text stands out from the rest of your message because each line of the original message is marked in some way, often with an arrow (>).

Wherever you look on the WWW, people and companies are providing their e-mail addresses. This has made communication with strangers easier than ever. If you read an interesting article on-line you can immediately send the author an e-mail and the chances are that you will receive a prompt reply.

Because e-mail is less formal than traditional correspondence, it has more of a conversational style than conventional letter writing. You are more likely to come across messages that start with 'Hi!' and end with 'Cheers' than messages which start with 'Dear …' and end with 'Yours sincerely'. E-mail messages also contain a high proportion of acronyms. Another feature, which is peculiar to the internet, is the use of emoticons, (sometimes called smileys). These can be thought of as an electronic form of body language and help you to put across the true meaning of your words; see figure 2 for an example.

Common Acronyms:

IMHO   In my humble opinion
BTW   By the way
FYI   For your information
TIA   Thanks in advance
ROTLF   Rolls on the floor laughing
RTFM   Read the friendly manual (The 'F' may be substituted for a strong swear word!)

Common emoticons, (viewed by turning your head sideways):

:-)   Happy
;-(Unhappy
;-)   Winking
8-D   Laughing (and wearing glasses)
:-!   Foot in mouth
:-&   Tongue-tied
:-X   Lips are sealed

E-mail programs

When you first open an account with an internet access provider, you are normally supplied with an e-mail software program. This will either be their own bespoke e-mail program which is often built into the rest of their software or a stand-alone third-party e-mail pro-

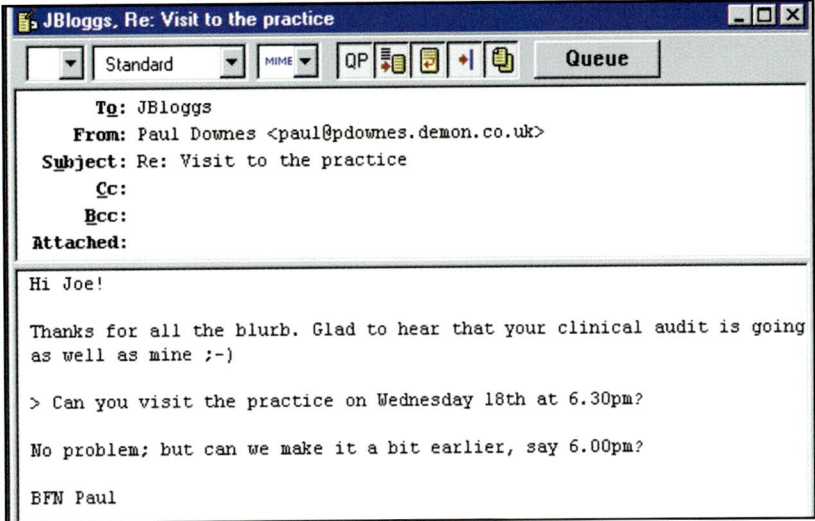

gram. You may also be able to send and receive e-mail using your web browser, depending upon which version of browser you are using. You can either continue to use the e-mail program that you are given by your provider, or you may later decide to try a different program if it offers you greater benefits.

The most popular third-party e-mail programs are *Eudora*[3] and *Pegasus Mail.*[4] In the next part of this series, I will use *Eudora Light 3.0* to show the practicalities of using an e-mail program; the process is similar for most other e-mail programs. (Check with your service provider that they support a POP3 mail service otherwise you will not be able to run *Eudora*; see the section about POP3).

Stand alone e-mail programs are often included on cover-mounted CD-ROMs given away with computer magazines, or can be downloaded directly from the internet.[3,4] It is common for e-mail programs to be available in two versions; a 'light' shareware version and a more advanced 'pro' version.

Shareware

A category of software that users can try out for a specified period of time so they can evaluate it before buying; therefore sometimes called an evaluation copy. This is an extremely popular method of distributing software through the internet. The cost of most shareware is cheap compared to commercial software.

Freeware

Software that is copyrighted by the author but made available to end users without charge.

Fig. 2 A short e-mail message showing the use of an emoticon, quoted text and an acronym

1. Survey on the use of the Internet by UK dentists:
   http://www.pdownes.demon.co.uk/survey.html
2. The Internet Bookshop:
   http://www.bookshop.co.uk/
3. *Eudora:* http://www.eudora.com
4. *Pegasus Mail Europe:*
   http://www.let.rug.nl/pegasus/

E-mail: how do you use the program?

A recent survey looked at how dentists in the UK used the internet.[1] It was found that the 73 respondents used 15 different e-mail programs between them. These programs can be divided into three main categories:

- The e-mail program supplied by the user's internet provider eg the *Mail Centre* that is built into *CompuServe's* software (fig. 1).
- The e-mail program that is part of the user's web browser eg *Microsoft Mail* or *Outlook Express* (fig. 2) from *Microsoft Internet Explorer*
- A stand-alone e-mail program such as *Eudora Light* (fig. 3) or *Pegasus*.

By comparing figures 1 to 3 you will see that there are many similarities between the different programs. I would recommend that those new to the internet should either use the e-mail program supplied by their internet provider or, if a program is not supplied, then the one linked to their web browser. If you later felt that you would benefit from additional features found in some of the stand-alone e-mail programs then you can easily select a different program to use as your e-mail client.

Configuring your e-mail program

Before you can start using an e-mail program, there are certain details that are needed to configure the software for your particular use:

- Your e-mail address
  eg **joe@jbloggs.demon.co.uk**, **mmouse@dial.pipex.com** or **ElmerFudd@msn.com**
- The address of your POP3 account; this address tells your e-mail program where to go to download your incoming mail. It looks like your e-mail address, but it often has an extra word after the @ symbol
  eg **jbloggs@pop3.demon.co.uk** or **mmouse@pop.dial.pipex.com**
- The SMTP address; this tells your e-mail program the address of the computer where it initially sends your outgoing mail
  eg **post.demon.co.uk** (*Note* that there is no @ in the SMTP setting.)

If you are using the e-mail program supplied by your internet provider, then you may find that some of these details may already have been entered into the e-mail program for you. Some internet providers use SMTP to both send and receive e-mail messages. If necessary, check the *Help* files of your particular e-mail program for further information on configuration.

Sending your first e-mail message

I will use *Eudora Light*[2] to illustrate how to send and receive e-mail messages. All e-mail programs work in a very similar way and so many of the instructions can easily be applied to other programs.

Start *Eudora* but do not yet connect to the internet; reduce your telephone bill by writing your mail off-line. Click on the ○ **New Message** ○ icon and a message window will appear on the right hand side, see figure 4. The upper half of the window has headings for you to enter information about the e-mail, while the blank bottom half of the window is where you type the text of your message. The various headings are:

To:
This is where you type in the e-mail address of the person who is to receive the message. If you want to send the same message to a second person, enter a second e-mail address, separated from the first address by a comma.

E-mail addresses can look very confusing to the uninitiated, but they all follow a similar pattern. They normally have the form 'something@name of provider' where the 'something' bit is usually the same as your account name. The customary naming convention is to use either your full name or initial plus surname, (without any spaces). For example:

bugsbunny@dial.pipex.com or **DuckD@msn.com**

My e-mail address is:
paul@pdownes.demon.co.uk
(pronounced 'paul **at** pdownes **dot** demon **dot** co **dot** uk').

Many providers allow up to five e-mail addresses with one subscription. Therefore different people within a dental practice, or individual members of a family could each receive their own private e-mail at separate e-mail addresses on the same computer, for example:

karien@pdownes.demon.co.uk and **jeremy@pdownes.demon.co.uk**

The form of e-mail address used by most CompuServe members is a bit trickier since you need to convert their address before you can send them a message on the internet. Although some members of CompuServe do have a personalised e-mail address, most have an address which consists of two numbers separated by a comma. To convert their

---

Part 4 of this series explains how a typical e-mail program works. The various features of e-mail programs are discussed and I will show you how to join a mailing list.

Key:

| Text that is of a more technical nature

password
Word(s) typed in at the keyboard

web browser
Keyword defined in the margin

readme.txt
The name of a file (or document)

○ **Forward** ○
The name of a button (or icon) in a program

File, Open
Step-by-step instructions using the menu bar

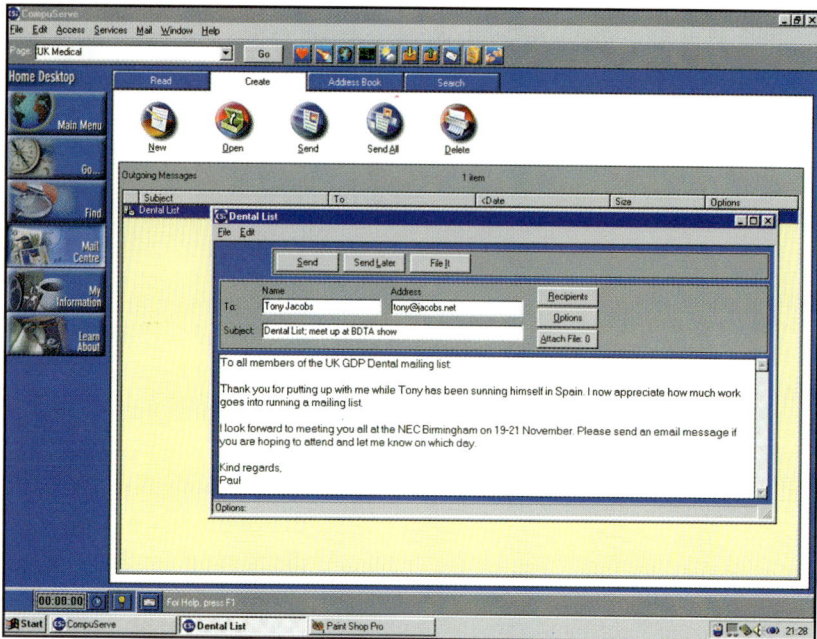

Fig. 1 Screen shot showing 'Mail Centre', the e-mail program that is integrated into CompuServe's software (Reprinted by permission of Compuserve)

Fig. 2 Screen shot showing the e-mail program *Outlook Express*. It is included with the free *Microsoft Internet Explorer 4* web browser. Subscribers to Pipex use the same program, except that the Pipex logo replaces the Explorer logo (Reprinted by permission of Microsoft Corporation)

address, change the 'comma' for a 'dot', and add **@compuserve.com**; for example 100123,1234 becomes **100123.1234@ compuserve.com**.

Finding someone's e-mail address

E-mail addresses are now commonly seen on people's headed notepaper and business cards to supplement their postal address and telephone number. Unfortunately, unlike postal addresses and telephone numbers, a printed directory of e-mail addresses does not exist. (The same is true for fax and mobile telephone numbers). If you know that the company or institution that they are associated with has a site on the WWW, then that can be a good place to start your search.

There are several electronic directories[3,4] on the WWW where you may be able to find someone's e-mail address by carrying out a search on their full name; however I have had

only limited success with this method. At the end of the day, by far the easiest way to find someone's e-mail address is to write or telephone them and ask them for it!

From:
This is your own e-mail address and it is automatically added to all your outgoing messages.

Subject:
The subject line allows you to give the recipient of your message an idea of what the message is about. eg 'Re: Restorative meeting at FGDP'.

Cc:
Type the e-mail address of any person whom you wish to receive a carbon copy of the message. The other recipients of this message will know that this copy has been sent.

Bcc:
Type the e-mail address of any person whom you wish to receive a blind carbon copy of the message. The other recipients of this message will not know that this copy has been sent.

Attached:
You can send files, (for example pictures, sounds, a Word document or Excel spreadsheet), attached to your e-mail message. Click on the ○ **Attach File** ○ button, (it looks like an envelope and letter held together by a paper clip), to open up a dialogue box which enables you to select the file you wish to attach. The name and location of the file will then appear under the attached heading. There is a toggle button where you can choose between coding the attached file by MIME or BinHex; choose MIME since this method is used by most e-mail programs.

MIME, BinHex and UUEncoding

E-mail was designed to send only basic text codes, (ASCII 7-bit characters), between networked computers, (that is why the £ sign does not appear correctly when sent in an e-mail message). Files such as graphics (*.bmp*), sound (*.wav*) or Word documents (*.doc*) contain more fancy coding than just basic text. Therefore files need to be converted, (or coded), into basic text codes before they can be attached to e-mail and sent

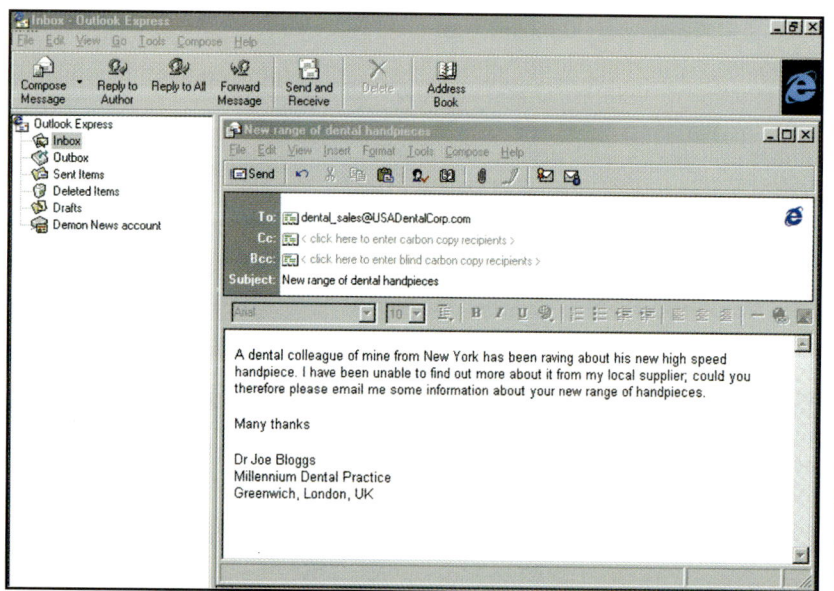

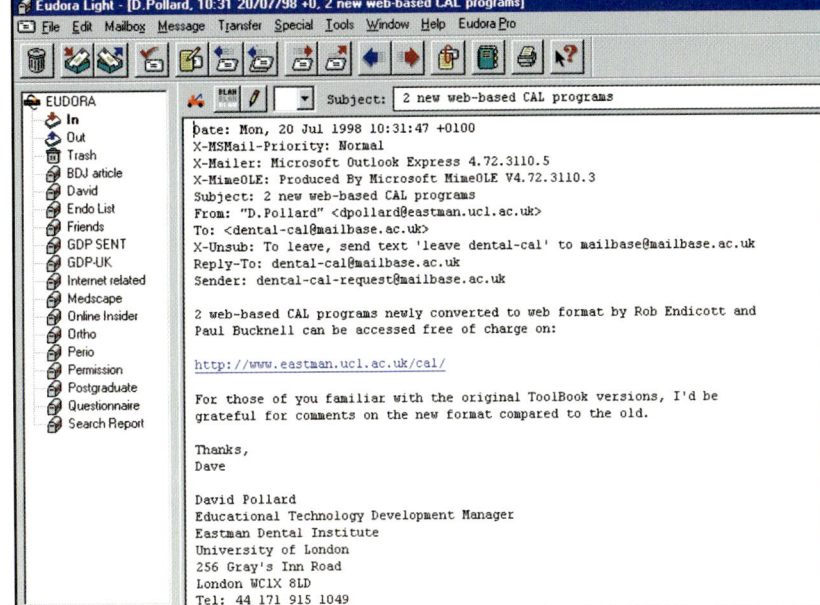

over the internet. The most popular methods for doing this are called Multipurpose Internet Mail Extension (MIME), UUEncoding and BinHex. MIME differs from the conventional UUencoding/decoding in that once having received a MIME message the relevant application associated to the MIME file will automatically be initiated. BinHex is mainly used by Apple Macintosh computers.

Naturally enough, not only the sender but also the recipient of a message containing coded files must have an e-mail software package that is able to handle the appropriate format so that it can be decoded at the recipient's end. The best e-mail programs code and decode the attached files automatically; others require you to manually run a separate coding/decoding program such as *WinUUE*.[5] If you have a copy of *WinZip*[6] you can use this to open MIME, BinHex and UUEncoded files and extract their contents.

You are now ready to type your message in the bottom half of the window. You can choose to have a 'signature' added to the end of every message you send. This could contain other details about yourself, such as your work address and telephone number, or a description of what you do or where to find your home page on the WWW. You set up the signature by selecting **Signature** from the **Window** menu and typing the text you would like in the window.

Signatures

A signature is a way of automatically specifying what appears at the bottom of every e-mail that you send. As a general rule, it is considered impolite to have a signature which is longer than four lines. It is also frowned upon to decorate your signature with ACSII art (cartoon characters made from keyboard characters); modern e-mail programs often display the text of the message in a proportional font rather than the old monospaced font, resulting in something which is impossible to recognise. Some people also feel that, for reasons of security, it is not wise to advertise your home address and telephone number with your e-mail messages. Most e-mail programs allow you to have more than one signature file.

Headers

Every e-mail you receive has a header that tells you who the message is from and the path it has travelled to reach you. Most e-mail programs normally give you the option to hide or show the header.

When you have finished typing your message, click the ❍ **Queue** ❍ button. The message will now be queued in the ❍ **Out** ❍ box ready for when you next go on-line. You can continue to write and queue further messages if so desired.

To send your message(s), connect to the internet and start *Eudora*. You can configure *Eudora* to send your messages and check for incoming e-mail automatically. The other option is to send and retrieve your e-mail manually(select **File, Send Queued Messages** and **Check Mail** from the menu bar).

Retrieving and reading your e-mail

Before you can retrieve your mail you will be asked to type in your password. This is normally the same as your password for your internet connection. *Eudora* stores all incoming e-mail in the ❍ **In** ❍ mailbox . Double click on the ❍ **In** ❍ mailbox and a window opens showing a summary of all your incoming messages. They are listed in the order they were received, with the most recent message listed last. Unread messages are designated by a bullet in the Status column. Double-click on any message summary to read the full message.

Responding to an e-mail

It is very easy to respond to an e-mail message since you do not have to type in the person's e-mail address or the subject line and you can quote all or part of their text in your reply. With their message open, just select **Reply** from the **Message** menu. A window will open with the To:, From: and Subject: lines automatically

Fig. 3 Screen shot shows the e-mail program Eudora Light. The main window shows an e-mail message that has been received and is ready to read. (Figs 1 and 2 show the composing of new messages) (Reprinted by permission of QUALCOMM Incorporated)

filled in and the original message copied into your reply and 'quoted' with a '>' sign at the start of every line. Select and delete any of the quoted text that you do not wish to appear in your reply.

Other features of *Eudora Light,* which should be found in any good e-mail program include:

Address book
It is possible to build up a useful 'book' of e-mail addresses by automatically creating entries from your incoming e-mail. Each entry can be for an individual person or you can create a list for a group of people, such as a committee or peer review group. The people whom you most often send e-mail can be put on a 'quick recipient list' so that their addresses are instantly available. Address books tend to be an integral part of your e-mail program; do not assume that you can transfer it easily to a new program.

Mailboxes and folders
In addition to the ○ *In* ○ and ○ *Out* ○ mailboxes, you can create your own personalised mailboxes in which to store your messages once you have sent or received them. You can also create a folder in which to store related mailboxes; for example a folder called 'Friends' may have a mailbox for each friend with whom you regularly communicate, plus an additional mailbox for friends whom you only occasionally keep in touch with by e-mail.

Filtering of messages
A useful management feature of e-mail is the ability to automatically store your incoming messages into the relevant mailbox. This can be done by a process known as filtering. For example, all incoming messages which match specific criteria, such as from a particular person or organisation can be copied to a certain mailbox.

Find text
If you cannot remember where you stored a particular e-mail message, you can carry out a fast and easy search on a word or phrase which you know is contained in the text of the e-mail.

Pretty good privacy
Pretty Good Privacy, (PGP), is a public-key encryption program.[7] Each user has two complimentary keys; a public key which you give to anyone you wish to communicate with and a private key which you keep secret. If someone wants to send you a message, they encrypt it with your public key. You, and only you, can then decrypt it with your private key. When you send out a message, you 'sign' it using your private key. The recipient decodes your 'signature' with their public key, verifying that only you could have sent it. If you want to send them an encrypted message you use their public key.

Formatted text
Newer e-mail programs allow formatted text, such as bold, italic and different fonts. Avoid this unless you know that the recipient will be using software with the same capabilities. It is also possible to include active hypertext links to WWW pages (URLs) with your e-mail message. If the recipient of the e-mail wishes to view the quoted page with their WWW browser then they simply double-click on the link from their e-mail program.

How to join a mailing list
Mailing lists were described in the previous part of this series; most are used as discussion groups, where subscribers to that list receives all the messages sent to the list, usually on a daily basis.

Joining a mailing list normally involves sending an e-mail to a central computer, where your subscription is then automatically processed by some special computer software. It is important to put exactly the right information in your e-mail otherwise the program at the other end will not understand your message and the subscription will fail. Automatic mailing lists are easy to spot since many either have the word 'majordomo' or 'listserv' in the e-mail address. Here are some guidelines on what information you should put :

1. Enter the address of the mailing list in the **To:** line.
2. Leave the **subject** line blank.
3. In the **body of the message,** type the subscription command line exactly as specified, (normally subscribe (List-Name) Your-First-Name Your-Last-Name).
4. Disable your **signature file.**

For example, if I wanted to subscribe to the mailing list that discusses orthodontics, I would

Uniform resource locator (URL) = the internet way of indicating the unique address of a particular page on the world-wide web

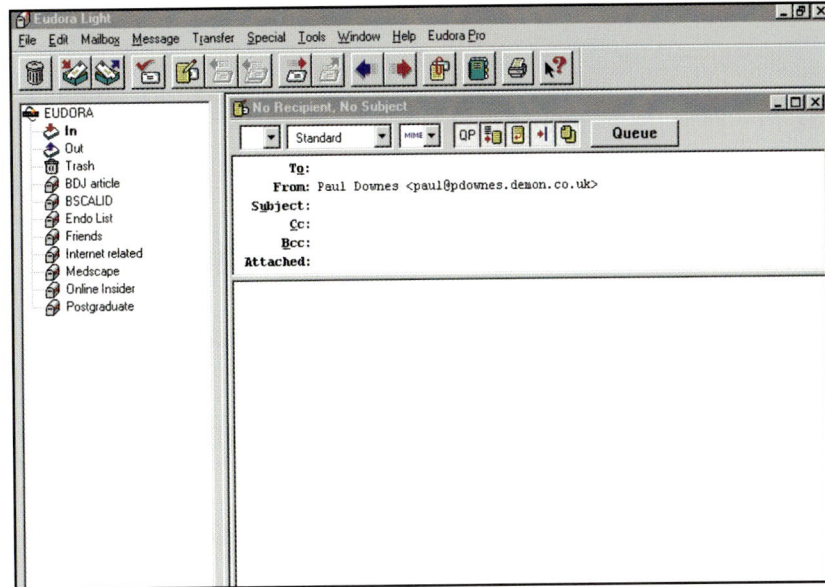

send an e-mail to **listproc@usa.edu** with the first line of the text as 'subscribe ORTHOD-L Paul Downes' (without the quotation marks).

Some mailing list subscriptions are managed manually, particularly medical/dental lists where the information you provide is first verified, since some lists are not open to the general public. With these sorts of lists, the layout of your e-mail message is not as critical.

For example, to subscribe to the UK GDP mailing list, send an e-mail to Tony Jacobs at **tony@jacobs.net** and include details such as your name, qualifications and where you work. Some examples of useful dental mailing lists can be found in Part 8 of this series.

Once you have subscribed to a mailing list, you are normally sent an e-mail a few days later with full details of how the list works; print this out and keep it somewhere safe. There is a feature whereby you can temporarily pause incoming messages from the mailing list. This is very useful if you subscribe to a busy list and you are going on holiday for a couple of weeks; it avoids you having to download hundreds of messages on your return.

There are other ways of using the internet to take part in either one-to-one or group discussions; some can be conducted partly off-line, (as with e-mail and mailing lists), while others require you to be on-line for the duration of the discussion. Here are some examples:

• *Newsgroups.* These will be discussed in the next part of the series.
• *On-line Service Discussion Forums.* CompuServe has its own forums which are for CompuServe members only; the UK Dental forum is very active and contains some very useful discussions. Click the ○ **Go** ○ button and type **UKMedical**; join the forum and then click on the ○ **Message Board** ○ button and look under item 6, Dental.
• *WWW Discussion Groups.* Both DERWeb[8] and Dentanet[9] host dental discussion groups from their web sites; the topics vary from

informationt technology to dental politics. Dentanet also provide a private discussion area for users of TMS dental software.
• *Live chat using Internet Relay Chat, (IRC).* IRC is a protocol for communication between computers. You install an 'IRC client', which is special software[10] that enables you to take part in live discussions. You first connect to an 'IRC server', which is a type of computer on the internet which transmits all the chatting around the world. You then join a channel, which is a bit like a room where people chat about one particular topic. You can read all the chat taking place and when you type a message, it instantly appears on the screens of the other people who are tuned into the same channel.
• *Live chat using a web browser.* DERWeb's WebBoard not only features a discussion forum but also a real-time chat area.
• *ICQ (pronounced 'I seek you').* This is an internet tool[11] that works in the background while you are using other internet applications. It informs you when friends, (who are also ICQ users), are on-line, enabling you to contact them at will. In the same way, when you log onto the internet, ICQ automatically announces your presence to your friends, (you can easily disable this announcement if you do not want to be disturbed). You can chat, send messages and files, and even play games. ICQ currently has more than 15 million subscribers. (The on-line service AOL has a similar system called 'buddies' but this is limited to members of AOL).

Fig. 4 The screen shot shows the e-mail program *Eudora Light*; the various mailboxes can be seen in the window on the left hand side, while the message window on the right is ready for a new message to addressed and typed (Reprinted by permission of QUALCOMM Incorporated)

Web browser = software program that allows you to view pages on the world-wide web

1 Survey on the use of the internet by UK dentists: http://www.pdownes.demon.co.uk/survey.html
2 Eudora: http://www.eudora.com/
3 Four 11: http://www.four11.com/
4 Bigfoot: http://www1.bigfoot.com/
5 WinUUE: http://www.neosoft.com/~pane/
6 WinZip: http://www.winzip.com/
7 Pretty Good Privacy: http://www.nai.com/default_pgp.asp
8 Derweb: http://www.derweb.ac.uk/
9 Dentanet: http://www.dentanet.org.uk/
10 Stroud's Consummate Internet Apps List: http://cws.internet.com/
11 ICQ: http://www.icq.com/

5 Newsgroups/File Transfer Protocol

Newsgroups (USENET)

When you need to find the answer to a particular question, then you will discover that newsgroups can be one of the most helpful parts of the internet. However, when you hear some bad publicity about the internet, it invariably concerns pictures or articles found in newsgroups. The main worry is how easy it is to gain access to information on sensitive topics, such as child pornography or practical advice on euthanasia. Luckily, this type of subject matter only constitutes a fraction of the material available from newsgroups. The internet has often been likened to a large city; most of it is perfectly safe but you and your family should always exercise caution when wandering around dubious areas.

What are newsgroups?

Newsgroups are a type of electronic bulletin board where people with a particular interest or hobby can post questions, answers and information that can be of great help to other like-minded people. There are more than 23 000 different groups to which you can subscribe and the number of articles per day have gone from 30 000 in 1993 to almost 250 000 in 1997. (The words 'newsgroup' and 'subscribe' are misnomers since the one thing that newsgroups contain very little of is actual news and it costs nothing to subscribe to a group).

What do I need to enable me to use newsgroups?

1. Access to a news server; this is a computer that distributes the messages sent to newsgroups. Most UK internet providers provide this access as part of their membership package. The range of groups available to subscribe to varies with each internet provider.
2. Newsreader software; this is a program that reads and sends newsgroup messages. There are lots of different newsreader programs

Table 1. Major newsgroup categories

alt	Alternative; anything goes!
comp	Computers; anything to do with computers
news	News; discussion about Usenet itself (NOT current affairs)
rec	Recreation; leisure activities such as music, films and sport
sci	Science; scientific discussion which includes medicine and dentistry
talk	Talk; discussion on controversial issues
soc	Social; social and cultural topics
uk	United Kingdom; topics related to the UK

available, and as with e-mail, the way each one works is very similar. You may use:

- The program that is an integral part of the software provided by your internet provider, eg the newsreader supplied by CompuServe (fig. 1).
- The newsreader that is linked to your web browser. Depending on which browser and which version you are using, this could be Netscape News, Netscape Message Centre, Microsoft Internet News or Microsoft Outlook Express (fig. 2).
- A stand-alone newsreader program such as the very popular *Free Agent*[1] (freeware) or the more powerful *Agent* (costing US$29 for the software or US$40 inclusive of disks and manual).

How do newsgroups work?

There are about 50 000 news servers around the world, exchanging information with each other on an almost continuous basis. When you send a message to your news server, that message is progressively copied from machine to machine, until it is eventually spread world-wide. It is estimated that about 500 Mb of articles are posted to newsgroups everyday, and because of this, the people who are in charge of news servers are selective about which newsgroups they take a feed on. For example, you are unlikely to find many Japanese newsgroups on your local news server. Some news servers will also censor out some of the more distasteful groups.

Configuring your newsreader

There are three main pieces of information that you need to type into your newsreader's configuration:

1. The address of your provider's news server, which normally takes the form of your provider's domain name, preceded by the word 'news' eg **news.demon.co.uk** (most internet providers preconfigure this for you).
2. Your e-mail address.
3. The name that you would like to appear in the From: field whenever you send a message eg **Joe Bloggs**

What newsgroup topics are there?

There are more than 23 000 different groups, (discussion forums), which means that they cover almost every topic imaginable; there is bound to be something of interest to everyone.

In this section, two very different internet applications are described: newsgroups — the world's biggest discussion forum; and file transfer protocol — the world's largest software library.

Key:

Text that is of a more technical nature

password

Word(s) typed in at the keyboard

web browser

Keyword defined in the margin

readme.txt

The name of a file (or document)

○ *Forward* ○

The name of a button (or icon) in a program

File, Open

Step-by-step instructions using the menu bar

the internet guide for dentistry

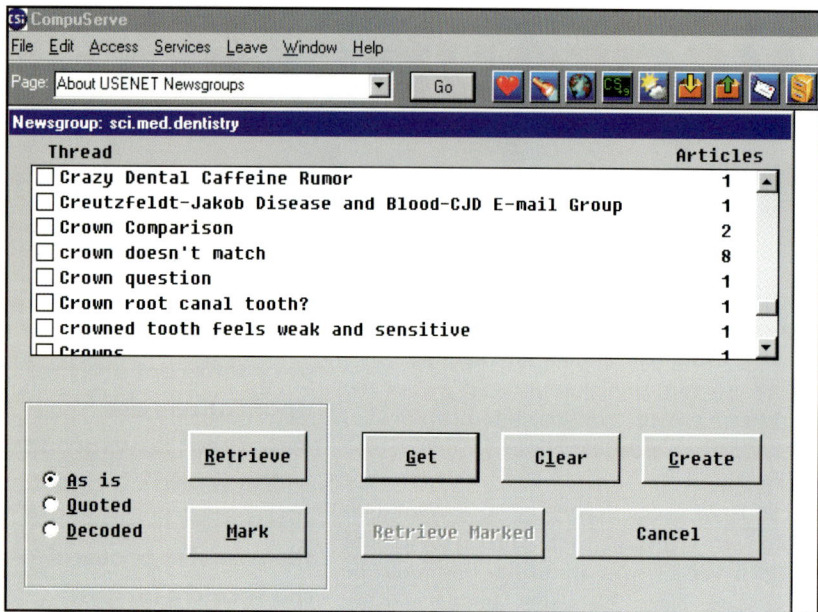

Fig. 1 Screen shot of the newsreader program that is an integral part of the software provided by Compuserve. (Reprinted with the permission of Compuserve)

You can subscribe to as many groups as you want; you can easily unsubscribe at a later date.

Newsgroups are divided into main groups by using a simple hierarchical naming system. The main topic can be identified by the first part of the newsgroup name. Sub topics follow the main heading and are seperated by a dot eg rec.music.classical.recording is a newsgroup that discusses classical music recordings and is part of the rec(reation) category. The major categories are shown in Table 1.

Using newsgroups; a practical example

All newsreader programs work in a similar way. I will show you how to access newsgroups, illustrating this process using the *Free Agent* newsreader program. To make this a practical and useful exercise, I will show you how to subscribe to the main dental newsgroup, sci.med.dentistry, and retrieve some articles from that group.

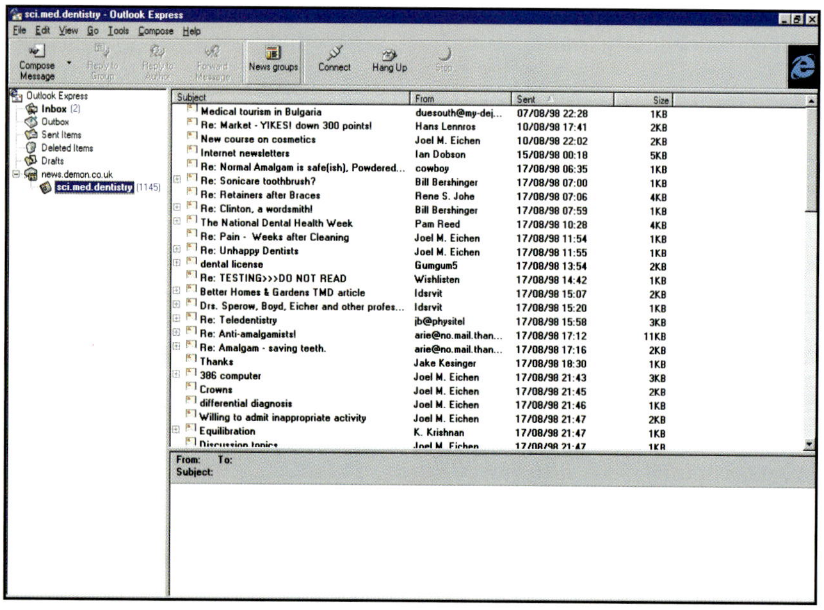

Fig. 2 Screen shot showing the newsreader program that makes up part of *Outlook Express*; it is supplied free with Microsoft Internet Explorer 4 (Reprinted with the permission of Microsoft Corporation)

When you start *Free Agent* you will see that it is composed of three panes: the Group pane, the Article pane, and the Body pane (fig. 3). Above the panes is the menu bar, (consisting of text drop-down boxes), and the tool bar, (consisting of graphical icons). In the following instructions I will be using the menu bar, but as with most Windows programs, many of the tasks from the menu bar can also be achieved by clicking one of the icons from the tool bar or by using keyboard shortcuts.

Since most dentists do not have a continuous connection to the internet, check that *Free Agent* is set up for off-line operation. You can do this by selecting **Preferences** from the Options menu and then clicking the ○ **Use Offline defaults** ○ button from the **On-line operation** tab.

1. Connect to the internet. Start *Free Agent* and download a list of all the available newsgroups held on your provider's news server. Select **Refresh Group List** from the **On-line** menu. You will see a progress message at the bottom of the screen showing you the course of the downloading process. Go and make yourself a cup of coffee since it will take 5–15 minutes to download the full list of several thousand groups, (luckily you only have to do this operation once). Once the groups are retrieved, they will be displayed in the Group Pane. Go off-line and have a look at what groups are available; they are listed alphabetically.

2. To subscribe to the group sci.med.dentistry, highlight its name in the Group Pane and select **Subscribe** from the **Group** menu.

3. Go on-line and retrieve the headers for your subscribed group by choosing **Get New Headers in Subscribed Groups** from the **On-line** menu. The headers will be displayed in the Article pane; they appear as a single line of text and give you a brief description of the article plus the name of the author. Once all the headers have been received, go off-line.

4. The next stage is to select which articles you are interested in reading, so that you can then download the body, or full text, of the articles. Read through the article headers and highlight the appropriate header from the Article pane and select **Mark** for retrieval from the **Article** menu. Repeat for each article you are interested in.

5. Go on-line. Select **Get Marked Article Bodies** from the On-line menu. Once all your article bodies have been retrieved, you can go off-line.

6. To read an article, highlight the article header in the Article pane and the full text or body of the article will appear in the Body pane, (fig. 3). If you wish to print the article, select **Print** from the **File** menu.

The method I have just described uses *Free Agent* as an 'off-line newsreader'; you keep down the cost of your telephone calls by doing a lot of the sorting and reading off-line, enabling you to retrieve the full text of articles that interest you. This is probably the best way to use *Free Agent* if you subscribe to a lot of busy groups. However, if you know that you will always want to read all of the articles in a particular group, you can set up *Free Agent* to automatically retrieve the bodies for all new articles at the same time as it retrieves the headers. To do this, right-click the selected group from the Group pane and then select **Properties for selected groups**. Click on the tab labelled **Retrieving** and tick the box **Retrieve bodies for all new articles**.

Other features of *Free Agent*, that are also found in other good newsreader programs include:

Posting and follow-ups. When you start a new message in a newsgroup it is called a 'posting'; a reply to an existing posting is called a 'follow-up'. (If you want to you can also reply directly to the original author by sending them an e-mail). To post a new article in *Free Agent* select **New Article** from the **Post** menu. To reply to an existing posting, highlight the relevant article and then select **Follow Up Article** from the **Post** menu.

Threading. A thread is a series of related articles grouped together. For example, when people post responses to a question, the question and all the responses are called a thread. In *Free Agent* the threads are grouped together and are shown in the status bar of the Group pane by a small icon with a plus sign. Clicking on this icon expands the whole thread so that you can see the original posting and all of the subsequent replies.

Signature files. As with most e-mail programs, it is possible to add different signatures to the end of your newsgroup postings. This is quite handy since you may want to use a different signature for postings to professional groups than you would for recreational groups.

Purging articles. If you never deleted any downloaded articles from your newsgroups, your newsreader program would soon become very cluttered. You can configure *Free Agent* to purge old articles after a set period. To do this, right-click any of your subscribed groups from the Group pane then select **Default properties for all groups**. Click on the tab labelled **What to purge** and select how much data you want to store on your hard disk from the available choices; this will automatically set the configuration for purging old articles.

Keeping articles. Occasionally you will come across an article that you do not want your newsreader to automatically delete. It is possible to mark the article so that it is always kept, no matter how you have configured your purge settings. In *Free Agent*, you highlight the article and choose **Keep Article** from the **Article** menu. A small padlock icon then appears next

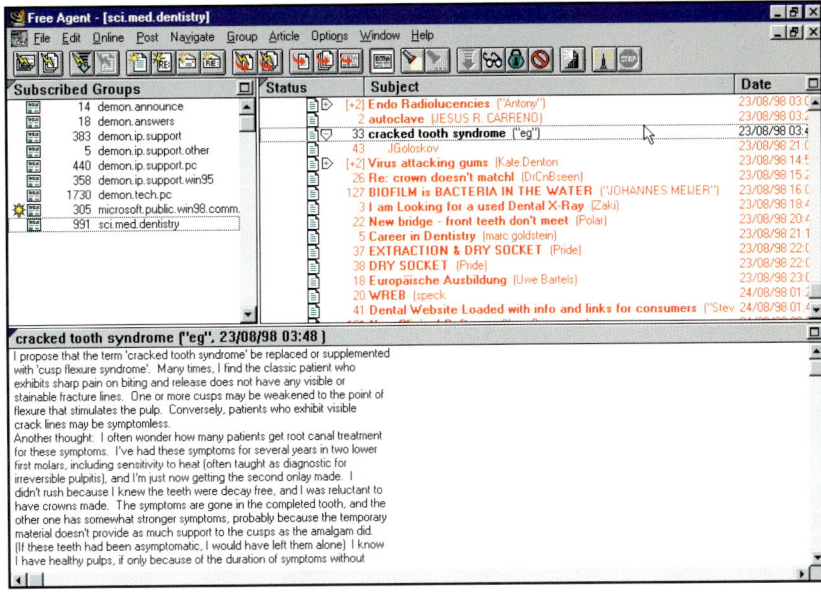

to the marked article.

Attaching files. You can attach files to your newsgroup postings, (such as images, sounds or programs), in the same way as you attach a file to an e-mail. However, because of the risk of downloading a virus, I would not recommend running any programs that are attached to a newsgroup article unless you are sure about the safe origins of the file.

Netiquette

Although there are no hard and fast rules about how to conduct yourself when posting a message to a newsgroup, there has developed a certain etiquette, (termed 'netiquette'), which I would strongly advise you to follow. Disregard these conventions at your peril! Some newsgroups seem to have a self-appointed group of people (termed 'Net cops') who take great delight in singling out anyone who does not obey these unwritten rules. Some of the netiquette guidelines are just as appropriate for using e-mail and they are all designed to make everyone's time on the internet easier and more efficient.

1. Prior to posting questions to any newsgroup, see if the group contains an article marked FAQ, (Frequently Asked Questions). Read this article; someone has put in a lot of hard work in compiling this valuable source of information and you may well find that it contains the answers to most of your questions.

2. You should read a newsgroup for about a week prior to posting to it; this is called 'lurking'. It gives you a good idea of the sort of things that are covered within the group; you may realise that there is a more appropriate group to which to pose your question.

3. As with e-mail, keep your signature to a maximum of four lines of text.

4. Do not send the same message to lots of groups at once. On a small scale, this is called 'cross-posting' and is just about tolerated. On a large scale it is called 'spamming' and will

Fig. 3 Screen shot showing the newsreader program *Free Agent*. It is made up of three main panes: the left-hand Group pane, the right-hand Article pane, and the bottom Body pane. The Group pane shows that nine groups have been subscribed to. The sci.med.dentistry group is highlighted, and the article headers for this group are shown in the Article pane. The body of the highlighted article is shown in the Body pane

Signature = a message that can be automatically added to the end of any e-mail or newsgroup posting. It may contain such things as your work address and telephone number

Web browser = software program that allows you to view pages on the world-wide web

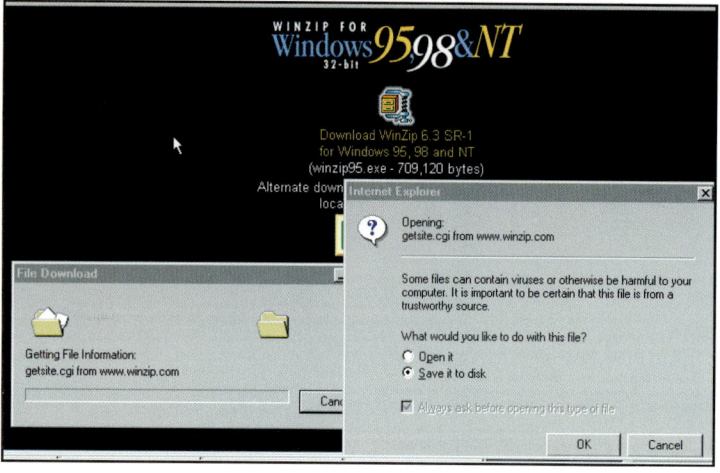

Fig. 4 Screen shot showing what happens when you use your web browser to download software, in this case, *WinZip*. Initially, The File Download window appears followed a few seconds laterby another window that asks you what you want to do with the file. Click on the ○ *Save it to disk* ○ option and then specify where on your hard disk you want to store the file

guarantee that you incur the full wrath of the Net cops.

5. Do not quote a long message and then simply type 'Yes, I agree' at the bottom.

6. Do not criticise others for their poor spelling and bad grammar, but try to make sure you use correct spelling and grammar yourself.

7. Do not type your entire message in uppercase because it makes it very difficult to read. IT IS CALLED 'SHOUTING'.

8. If you do have the misfortune to receive an abusive reply to one of your postings, think hard before you send an equally abusive retort. This type of message is called a 'flame' and by replying to it in a similar vein you will invariably start a 'flame war'.

Freeware = software that is copyrighted by the author but made available to end users without charge

File Transfer Protocol (FTP)

One of the marvellous things about the internet is the wealth of free or very cheap software avail-

Fig. 5 A screen shot from the FTP program *WS_FTP*. The left-hand pane shows the directories and files on your computer's hard disk. The right-hand pane shows the directories and files on the remote host computer to which you are connected. To download a file, first highlight it and then click the left-facing arrow situated between the two panes

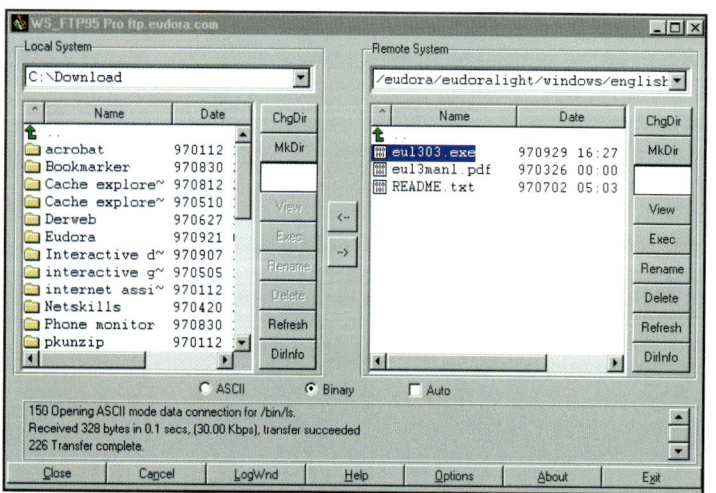

able to download directly onto your computer. This software could be an updated driver for your printer to make it work more efficiently, a dental computer-assisted learning program or some software to enhance your PC.

One very simple way of downloading software from the internet is to use your web browser. When you come across a page where you want to download some software, just click on the appropriate hypertext link, (this will normally be some text or graphic that contains the word 'Download'). Fig. 4 shows an example of this. If all you want to do is download small programs, say less than 2 Mb, then your web browser is all that you will require. However, most internet providers normally also include a dedicated File Transfer Protocol (FTP) program with the bundle of software provided when you take out an account.

Do I need a dedicated FTP program?

Dedicated FTP programs have been written to enable you to both download and upload files on the internet. You will need to use one of these if:

1. You want a faster and more reliable method than using your web browser to download large files.

2. You want to 'publish' your own home web pages on the internet, since you need to be able to securely upload your pages to your internet provider. You cannot do this using your web browser. However, members of AOL and CompuServe can use their service's built-in software for uploading pages to their server, and many web page editors such as Microsoft FrontPage include publishing features.

Logging onto FTP

When you connect to a remote computer using your FTP program, you are called the guest and the remote computer is called the host. The computers on the internet which store files for the general public to access normally also store private data for local use. Therefore, for authorisation purposes, passwords are used to control who can have access to which files. The way in which you log on and access the public archive of files is called anonymous FTP since it involves using the word 'anonymous' (or sometimes 'guest') when asked for your User ID. When asked for the password, just type in your e-mail address.

Using FTP; a practical example

I will explain how to use a popular freeware FTP program called *WS_FTP Limited Edition*,[2] and show you how to download a copy of the popular e-mail program, *Eudora Light*. (Even if you are using a different FTP program, the general principles will be the same).

In order to find a file you first need to know two things:

1. The name of the file you are looking for, (in this case *eul303.exe*).

2. The address of a host computer which has a copy of this file, (ftp.eudora.com), and the directory on which the file is stored, (*/eudora/eudoralight/windows/english*).

This sort of information is often quoted in Internet magazines, e-mail correspondence, newsgroups or WWW pages.

The appearance of an FTP program is similar to *File Manager* or *Explorer,* except that it simultaneously shows you the contents of your hard disk as well as the contents of the remote computer you are connected to by the internet. Downloading a file is simply a process of finding the file on the remote computer, highlighting the file and then instructing the host computer to download it to a chosen place on your computer's hard disk (fig. 5).

The stages involved are as follows:

1. Connect to the internet and start the *WS_FTP* program. If it is the first time you have used your FTP program, a Session Properties window will appear (fig. 6). If the session profile window does not appear, you can get to it by clicking the button labelled ○ **Connect** ○ at the bottom of the main program window.

2. Click on the tab labelled **General**. It is from this window that you set up the profile for your connection to a remote host computer. Once the profile information is entered, you can connect to a remote host by simply choosing the host from the **Profile Name** drop-down list and then clicking the ○ **OK** ○ button. The profile name can be anything since it is simply your description of the host FTP site. FTP programs normally come with a preconfigured list of the most popular FTP sites. Click on the drop down list to see which profile names have already been supplied. If the *Eudora* site is not present, add it now.

3. To add a new profile for *Eudora,* click the button labelled ○ **NEW** ○ and type in the details as they appear in figure 4. (Use your own e-mail address for the password).

4. Click the ○ **OK** ○ button. The main program window will now appear, showing the directory structure of your hard disk on the left and the directory structure of the remote FTP server on the right. The Message Log at the bottom of the window gives you details of the logging in and data transfer process.

5. Move to the directory on your hard disk where you want to copy the downloaded file. (I have created a directory on my hard disk called 'Download' and this is where I store files which I download from the internet). To move from a directory down to a sub-directory, highlight the desired directory icon and then click the ○ **ChgDir** ○ button, (to move up a directory, highlight the green arrow at the top of the directory file list and then click ○ **ChgDir** ○ button). Use this technique to work your way through the directory structure of the

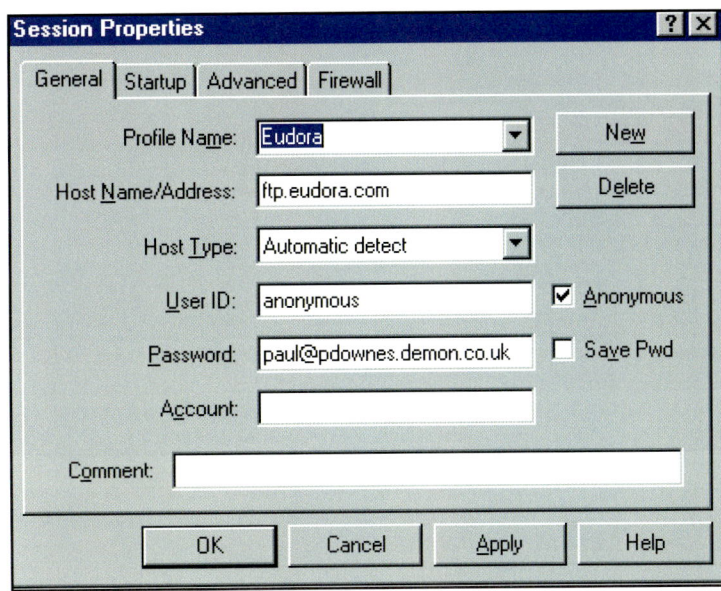

Fig. 6 The Session Properties window for the program *WS_FTP.* It is from here that you set up the different profiles for connecting to FTP remote sites. The settings for connecting to the Eudora FTP are shown

remote FTP server until you find the file *eul303.exe.* (You will need to move through the following sub-directories: eudora, eudoralight, windows and english).

6. The file can now be transferred to your computer by highlighting the file and clicking the left arrow button located between the two directory lists. A progress bar will appear, showing you the course of the download.

7. Exit the FTP program and go off-line.

8. To install *Eudora Light,* use *File Manager/ Explorer* to go to the directory where you downloaded the file. Locate the file *eul303.exe,* double click it and then follow the instructions from the installation program.

Zipped Files

The size of files for downloading through the internet are usually made smaller, (compressed), by special software; this helps to speed up the process of file transfer and reduce the time spent on-line. It is also useful for compressing program files because one compressed file can contain all the files necessary for the program to work. A common compression method is called 'zipping' and a 'zipped' file will have the extension *.zip*; for example *upgrade.zip.* Once you have retrieved the zipped file, you can 'unzip' it very easily using a program such as *WinZip* which is commonly available from internet magazine cover-mounted CDs or from the *WinZip* web site.[3] The program is free to use for evaluation purposes, but if you decide the program is useful and you want to keep it, you should pay $US29 to register the program.

Once you have installed *WinZip,* you can configure it to run a wizard every time you double-click on a zip file in *File Manager/ Explorer.* The wizard guides you through the process of unzipping the file and will even automatically install the software distributed by the zipped file.

Hyperlink = the links that tie together the world-wide web. Clicking on a link within a web page could take you to another page or even run an element of multimedia such as a video

1. *Agent* and *Free Agent*
   http://www.forteinc.com/agent/index.htm
2. *WS_FTP*
   http://csra.net/
3. *WinZip*
   http://www.winzip.com

Introducing the world-wide web

Although the terms world-wide web (WWW) and the internet are often used synonymously, they are actually two different things; the WWW is a subset of the internet.

The WWW is the world's largest store of easily accessible information; most of it is freely available. It is composed of millions of inter-linked documents, or 'pages', each of which can contain text, pictures, tables, sound, video or even virtual 3D-graphics. The pages are normally grouped together in 'sites'; for example, when people talk about the Denplan site,[1] they mean the collection of web pages produced by Denplan. Each site is normally, but not always, kept on one computer. Some computers will play host to many sites; for example, the DERWeb site[2] is stored on a computer, (a 'web server'), at the University of Sheffield which also hosts sites for other institutions such as the Confederation of Dental Employers (CODE) and the Dental Laboratories Association. Because of the ease of navigating from one page to another, it is not normally necessary to know on which computer the information is electronically held.

What makes the WWW special?

The WWW is quite different from more traditional sources of information such as books, magazines, television and video. There are many reasons for this:

- Web pages can be joined by 'hyperlinks' so that it is easy to jump from one page to another. The link may be between pages on the same site, but it is just as easy to click on a hyperlink and leap to a page on a web server in an entirely different country. A hyperlink normally takes the form of highlighted text, (sometimes underlined), or a graphic.
- Because of the way related documents are seamlessly joined by hyperlinks, it can make the process of finding further information on a particular topic almost effortless. For example, you may be looking at a page of information about orthodontics, and then by clicking on some text that says 'Associations' you jump straight to a page that contains a list of orthodontic associations with further hyperlinks to their home pages.
- It is possible to interact with web pages. Examples include a currency converter,[3] a loan calculator[4] and a virtual shopping basket.[5]
- It is easy to update the information held on a web page; for example, there are sites giving you current share prices[6] which are updated many times during the day.
- The web has tools called 'search engines' which enable you to search for a keyword, not just within a particular site, but on the entire contents of the WWW. The answer to the search is normally returned within a few seconds.
- You can 'publish' your own web site and the information it contains will become instantly available to anyone on the internet.

What do you need in order to use the WWW?

Apart from an internet connection, the only other thing you require to use the WWW is a software program which enables you to view the web pages. This program is called a web browser, and will be supplied by your internet provider as part of your membership package.

On-line services like CompuServe and AOL use their own customised browser which shows menus and content produced by themselves; the same browser is also used to show other web pages on the internet. Most ISPs provide you with a copy of a stand-alone web browser; either *Microsoft Internet Explorer (MSIE)* or *Netscape Navigator (NN)*; (figs 1 and 2).

Web browsers: which version should I use?

I would recommend that you use at least MSIE3 or NN3, which were released toward the end of 1996. At the end of 1997, Microsoft brought out Internet Explorer 4 and Netscape released Netscape Communicator. These are both integrated suites of internet applications that include version 4 of their web browser in addition to tools for scheduling, conferencing, e-mail, newsgroups, web page authoring and publishing. Both MSIE4 and NN4 can be downloaded free of charge from the internet,[7,8] and most internet access providers now supply one or the other as standard. They offer many enhancements over versions 3, but as with modern word processor programs, not all of these extra features will be used by the home user. They also require a more powerful computer to make the software run satisfactorily.

New features in MSIE4 include the ability to browse previously viewed web pages when

---

This part of the series introduces the world-wide web and attempts to explain how it works and what you need in order to use it. The use of a browser is also described.

Key:

Text that is of a more technical nature

password

Word(s) typed in at the keyboard

web browser

Keyword defined in the margin

○ *Forward* ○

The name of a button (or icon) in a program

File, Open

Step-by-step instructions using the menu bar

Return to home page

Hyperlink text

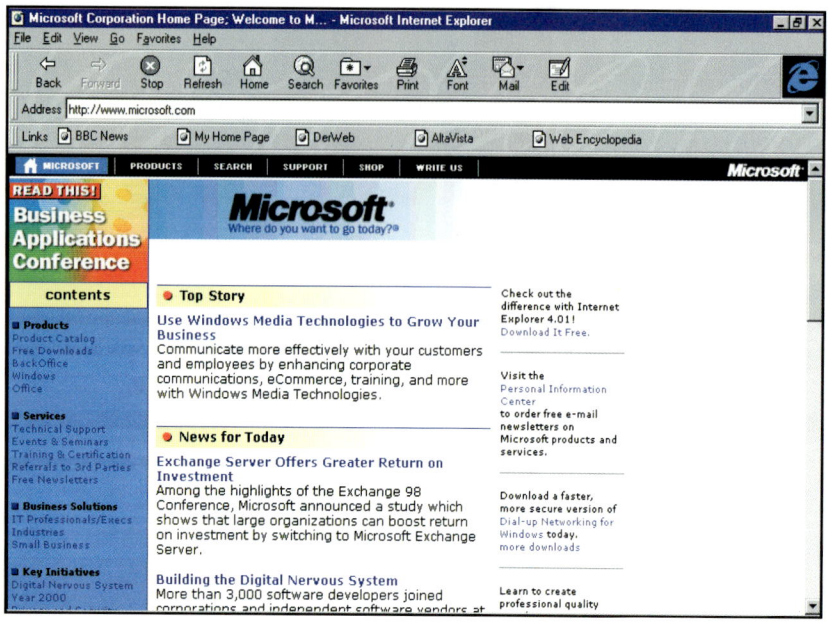

Fig. 1 Screen shot showing the web browser from Microsoft, called Internet Explorer version 3 (MSIE3). The downloaded page is the Microsoft home page (Reprinted with the permission of Microsoft Corporation)

Fig. 2 Screen shot showing the web browser from Netscape, called Netscape Navigator version 3 (NN3). The downloaded page is the Netscape home page (Reprinted with the permission of Netscape Communications Corporation)

you are off-line and 'subscriptions' which enable you to be automatically told when the content of pre-defined web pages change. Both MSIE4 and NN4 feature 'channels' which offer various areas of content such as news or sport and which automatically send you updated information without you having to request the data to be downloaded. These are perhaps more useful for people who use a continuous internet connection.

MSIE4 and Windows 95

If you are using Windows 95 and you install MSIE4, you are given the option of integrating web browser functionality directly into the Windows desktop. For example, Windows Explorer will look and behave just like a web browser with the addition of an address bar as well as favourites and back/forward buttons. This enables you to use the same method for finding files and programs on

your own computer as you use for exploring the WWW. In effect you are making changes to the Windows 95 operating system that take it one step closer to the new Windows 98 system.

MSIE4 and Windows 98

MSIE4 comes as an integral part of the Windows 98 software (a move that has landed Microsoft in front of the US Department of Justice to defend accusations of unfair business practice).

What is a URL?

A uniform resource locator (URL) is the internet's way of indicating the unique address of a particular page on the WWW so that your browser knows exactly where to go on the server to find a page. URLs are visible everywhere these days; tacked onto the credits of television programs, on commercials and magazine advertisements and even printed onto shopping carrier bags.

The full form for a URL looks like **http://www.tesco.co.uk/** or **http://www.bbc.co.uk/**, but these are often shortened to **www.classicfm.co.uk** or **www.barclays.co.uk**. These URLs will all take you to the home page or starting point of the company's web site; from there you can explore the other pages on the site, for example, **www.movies.co.uk/cinemas/london/plaza**. This fictitious URL looks a bit like a directory path — which is exactly what it is. It gives the exact location of the web page 'plaza' on the web server, (notice that it uses a forward slash (/) rather than the usual backwards slash (\) found in DOS and Windows).

How does the WWW work?

The web is based on a set of rules for exchanging text, images, sound, video, and other multimedia files, which is collectively known as hypertext transfer protocol (HTTP). Web pages can be exchanged over the internet because web browsers (which read the pages) and web servers (which store the pages) both understand HTTP.

Web pages are written with the same computer language; HyperText Markup Language (HTML). HTML is a subset of SGML and has evolved from the printing industry. HTML started as a simple form of tagging, or formatting text, and has developed to include commands for integrating the multimedia and interactive elements found on many web pages. HTML instructs your web browser how the text and graphics should appear (eg bold, italic, font size, centred etc.) as the page is downloaded onto your computer. Figures 3 and 4 show a simple example of what 'raw' HTML looks like, and then how it would appear when viewed by a web browser.

```
<BODY>
To make a word appear <B>Bold</B>, use the Bold tags.<P>
You can also make words appear in <I>Italics</I><P>
<H1>or with a first order heading</H1>
</BODY>
```

Fig. 3 This figure shows an example of raw HTML and how text can be formatted using tags. Notice how the tags, such as <B> are turned off using the /command (The tag for a new paragraph, <P>, does not need to be turned off)

To make a word appear **Bold**, use the Bold tags.

You can also make words appear in *Italics*

or with a first order heading

Fig. 4 How the HTML text from figure 3 appears when viewed using a web browser

The importance of HTML is the fact that it is not system-dependent; this means that your web browser will be able to view any page written in HTML no matter what operating system your computer is using (for example, IBM-PC or Apple Macintosh).

Using a web browser — a practical example

To illustrate how a WWW browser works, I will demonstrate some simple tasks using *MSIE3* and *NN3*; these were the browsers most commonly used by UK dentists who responded to a recent survey (32% and 30% respectively).[9] Figures 1 and 2 show that the layout of the two browsers are very similar. Each browser has a main menu bar of commands, a button bar with graphical buttons, an address box for typing the URL of the page you want to visit and some buttons which have been pre-configured to take you to a particular web site. The main area of the browser shows the web page currently being viewed.

I will show you how to access the DERWeb site (fig.5), one of the most useful sites for UK dentists. I will explain how to navigate around the site and introduce you to some common features of the WWW. (Bear in mind that the layout of the site is due to change toward the end of 1998.)

1. Connect to the internet and start your web browser. Your browser is normally pre-configured to show a default 'start page'; this is often either the home page of Microsoft or Netscape, or the home page of your internet provider.

2. In the browser's address box, type in the shortened version of the URL for DERWeb, which is **www.derweb.ac.uk**, and press the Enter key on the keyboard, (you do not need to type in the 'http://' part of the URL).

3. The DERWeb home page should start to load into the browser; first, the text will appear and then the graphics. The cursor

will change from a pointer into an hour-glass, and then back to a pointer once all of the page has downloaded. (Another indication that a page is downloading is that the *MSIE* or *NN* logo at the top right-hand corner of the browser will appear animated).

4. If the page is too big to fit onto the screen, a scroll bar will appear; this allows you to scroll up and down the page.

5. It would be handy to be able to come back to this site on another occasion and not to have to type in the DERWeb URL again. You can do this by getting the browser to remember the page's URL; try doing this now. In *MSIE*, click the ○ *Favorites* ○ button and then select **Add To Favorites** and click ○ **OK** ○. You can return to this page at any time in the future by clicking on the ○ *Favorites* ○ button and selecting the site from the drop down list. (*NN* has a similar feature called 'Bookmarks', accessible from the menu bar).

6. You can normally see whether any object on a page is a hyperlink by moving the mouse pointer over the item. If the cursor pointer changes to a hand, the object is a hyperlink and the URL of where it leads to will appear

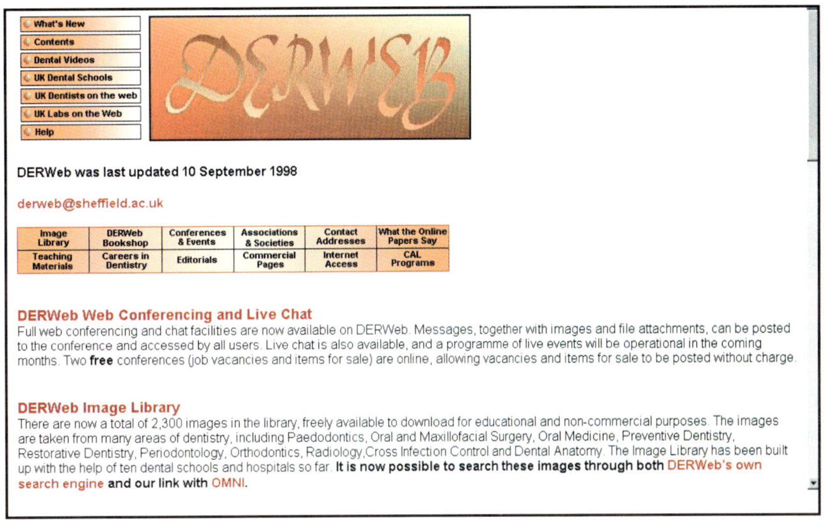

Fig. 5 The home page of DERWeb, showing the various hyperlinks to other sections of the web site. The content is updated on an almost daily basis and there are plans to change the layout of the site toward the end of 1998

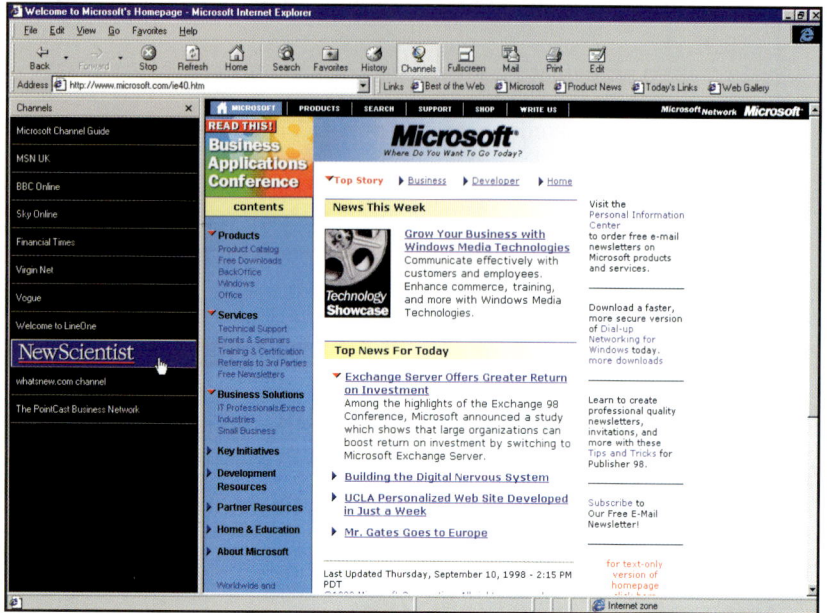

Fig. 6 This screen shot shows the Microsoft Internet Explorer 4 web browser (MSIE4). It illustrates the browser bar that appears to the left of the main window when clicking on the ○ *Search* ○, ○ *Favorites* ○, ○ *History* ○ or ○ *Channels* ○ button. The list in the browser bar controls the content of the main window (Reprinted with the permission of Microsoft Corporation)

Newsreader = software program used to follow on-line discussions in newsgroups

FTP = file transfer protocol; a protocol by which data is either downloaded to your computer or uploaded to another computer on the internet

at the bottom of the screen. Try clicking on a hyperlink; it will either take you to a place on the same page or to a completely different page either on the same site or another related site. As you move from page to page, and from web site to web site you will notice that the URL in the address box will automatically change.

7. Use the browser's ○ *Forward* ○ and ○ *Back* ○ buttons to retrace you steps. (The ○ *Back* ○ button can also be useful when you encounter the message 'page not found').

8. If you want to stop a page from downloading, click the ○ *Stop* ○ button.

9. Retrace your steps back to the DERWeb home page; you will find a set of icons on the bottom of most DERWeb pages and clicking the icon that resembles a house will take you straight back to DERWeb's home page. Now click on the hyperlink labelled derweb@ sheffield.ac.uk (this can be found near the top of the home page). Clicking this will automatically open your default e-mail program and start a new message with DERWeb's e-mail address already inserted. This feature is found on most web sites and is very useful if you wish to contact the author of a particular web page. On many web sites this sort of e-mail hyperlink is often labelled 'reply' or 'feedback'.

10. If a page does not appear correctly, for example a large graphic may sometimes stop downloading, click the ○ *Refresh* ○ button in *MSIE* or the ○ *Reload* ○ button in *NN*.

11. To return to your browser's default 'starting page', click the ○ *Home* ○ button.

12. The ○ *Print* ○ button will print the current web page.

13. In both *MSIE* and *NN*, you have the option of choosing whether or not any images on a

page are automatically downloaded. To turn off the autoloading of images in *MSIE*, click **View**, **Options** and under the **General** tab, remove the tick from the **Show pictures** box. If you are using *NN*, select **Options**, and remove the tick from **Auto Load Images**.

14. *NN* has an ○ *Open* ○ button, which functions as another way of typing in a URL and a ○ *Find* ○ button which allows you to search the current page for a particular word or phrase.

15. To jump directly to a previously-viewed page, click **Go** from the browser main menu to show a list of the sites that you have visited during your current session on-line. Click on the page that you want to go to.

16. You will find various pages on the DERWeb site that contain 'forms', for example the search boxes on the on-line bookshop and the image library. Forms take the shape of text entry boxes, buttons, selection lists and checkboxes; they enable the user to enter information into the current page. Forms allow interaction between the user and the WWW and is probably the most powerful feature of the internet, enabling two-way communication, search facilities and shopping.

17. From the home page you will also find links to the DERWeb Web Board and Live Chat pages. These are areas for general dental discussions, announcements, items for sale and job vacancies. They are accessible once you register your name and a password. There is a help page to guide you through the process of reading and replying to the discussions. You will often come across commercial web sites that require you to register before you are given access to information on certain parts of the site. Registered access is usually free; it is normally a matter of the owners of the site needing to have a record of who has visited the site and how often, so that they can satisfy their sponsors or advertisers.

18. Once you have registered and entered the Web Board you will notice that the page has been divided into three distinct areas or 'frames'. The narrow frame at the top of the page contains some toolbar buttons. Below this is a frame on the left that contains a list of the messages. Clicking on a message will show the content of that message in the frame on the right. The left-hand frame remains static to make navigation within the site easier and it controls the content of the right-hand frame.

Problems with frames

Many large web sites use frames to structure the content of the site. However, there are two occasions where you may wish to turn off the

frames option:
- Some browsers (mainly older versions) do not support the use of frames.
- If you are using a small screen, you may find that the content frame is annoyingly small.

Because of this, you will often find that the authors of the site provide some text or a button that you can click on that downloads a version of the page that does not use frames.

As you explore the web, you may notice that hypertext links you have visited before appear in one colour, (normally red), while those which you have not visited remain in another colour (normally blue). You can configure your browser as to how these visited links appear and after how many days they expire; In *MSIE* click on **View**, **Options**, and see the settings under the **General** and **Navigation** tabs. In *NN* click on **Options**, **General Preferences**, and see the settings under the **Appearance** tab.

Browsers take the raw data about a web page, interpret it and then display the page on your screen. This has two main repercussions:
- The appearance of a particular page may vary depending on which browser you are using.
- You can easily change the appearance of any existing page, for example, increase the font size of the text to make it easier to read. In *MSIE*, click on the ○ *Font* ○ button to change the font to one of five different sizes. In *NN* you can set both the font size and type by clicking on **Options**, **General Preferences** and looking under the ○ *Fonts* ○ tab.

One very important point to remember about the WWW is that its content is always changing. If you bookmark a page and return to it at a later date, do not be surprised if the page is missing or the content of the page has changed. Therefore, if you find some information that you want to keep, you should either print the page or save it to your hard disk by clicking **File**, **Save As File** or **File**, **Save As** from the menu. If you just want to save a graphic from the page to your hard disk, place the cursor over the image, and right click the mouse to bring up a new menu box. Select the command **Save Target As...** or **Save Image As....**

Copyright

Strictly speaking it is an infringement of copyright to save a web page as a file to your hard disk, (let alone print it), unless you have permission from the author. Saving and re-using images can also be a breach of copyright and you need to check the terms and conditions of each site before doing so. Having said that, your computer will automatically copy the contents of every web page you visit and store it in a 'cache' on your hard disk, (see Part 9 for a full description of a

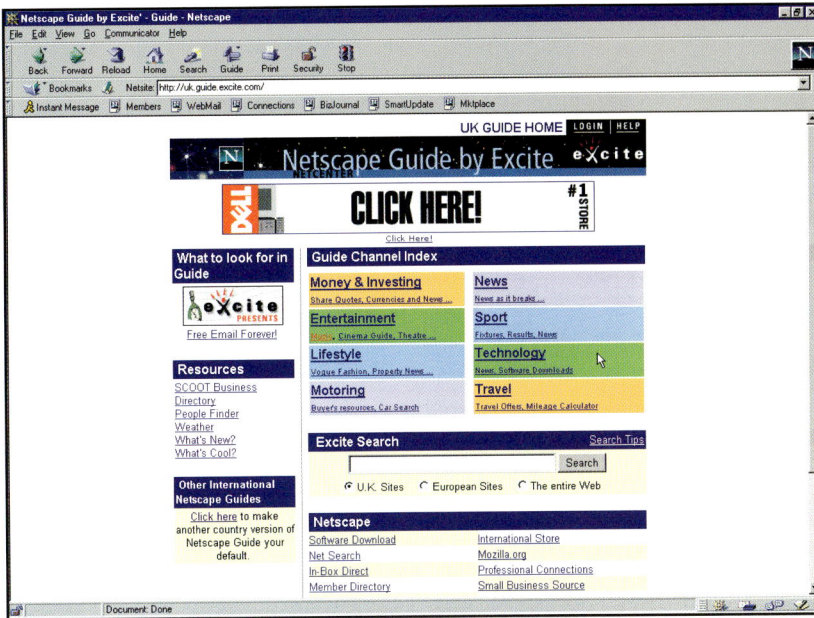

cache). *MSIE3* stores data from visited web pages in the *Windows\Temporary Internet Files* directory, while *NN3*, is likely to be store them in *Program Files\Netscape\Navigator\ Cache*. The World Intellectual Property Organization[10] have recently ruled that it is not an infringement of copyright for a cache to store web data.

MSIE3 and *NN3* have a built-in e-mail program and support FTP for the downloading of files; *NN3* also acts as a newsreader. These functions are easy to use, but may not provide all of the features available from specialised stand-alone e-mail and newsreader programs.

Remember that when you close down your web browser, you are still connected to the internet; close your internet connection in the normal way.

Additional buttons in MSIE4

Clicking on ○ *Search* ○, ○ *Favorites* ○ or the two new buttons ○ *History* ○ and ○ *Channels* ○ will bring up a browser bar as a separate window pane, to the left of the main browser window (fig. 6). As with Windows Explorer, you use lists which appear in the left pane to control what is displayed on the right. *MSIE4* also has a button for ○ *Full screen* ○ view and an ○ *Edit* ○ button which opens up the viewed page for editing (in my case it uses *Microsoft Word 97*).

Additional buttons in NN4

Two additional buttons appear along the main button bar of NN4. The ○ *Guide* ○ button links to a page that contains local information about news, sport and weather (fig. 7). Clicking on the ○ *Security* ○ button shows you the various security settings for your Netscape browser, e-mail and newsgroup programs.

1. Denplan:
   http://www.denplan.co.uk/
2. DERWeb:
   http://www.derweb.ac.uk/
3. Oanda Currency converter:
   http://www.oanda.com/
4. Loan calculator:
   http://mortgage-source.com/cambridge/mortpay.htm
5. The Internet Book Shop:
   http://www.bookshop.co.uk
6. Share prices:
   http://www.esi.co.uk/
7. Microsoft:
   http://www.microsoft.com/
8. Netscape:
   http://www.netscape.com/
9. Survey on the use of the internet by UK dentists
   http://www.pdownes.demon.co.uk/survey.html
10. World Intellectual Property Organization:
   http://www.wipo.int/

7 Successful web searching

In order to make the most of the internet, there are a number of things you need to do:

- Ensure that you have the right hardware and that you sign up with an internet access provider that suits your requirements, (covered in Part 2 of this series).
- Get to know how your internet software works and what it can do, (covered in Parts 3,4,5 and 6).
- Learn how to use the internet efficiently; cutting down on wasted time and money (covered in Part 9).

There is one thing missing from this list and that is knowing how to find a particular item of information from the vast amount of data stored on the internet. Part 8 of the series will guide you to some of the best dental resources on the internet, while this section will describe the tools you can use to track down the information you are looking for. 'Successful searching' means the ability to find that illusive piece of information quickly.

Methods of finding information on the internet

There are three fundamental techniques used for finding information on the internet:

- Free-text search using keywords.
- Directories of information, divided into various subject topics.
- Hyperlinks.

Using these three electronic techniques, a wide range of tools have been developed to help with the process of finding specific information on the internet:

- Major search engines eg AltaVista[1]
- Major directory services eg Yahoo![2]
- Search engine software for finding information within a web site
- Web sites that allow you to search a large database of information by the internet eg MEDLINE by PubMed[3]
- Gateways to evaluated material eg OMNI[4] for biomedical resources.
- Hyperlinks from individual web sites, newsgroup articles and e-mail messages.

You do not require any extra software to use these tools since they are all accessed from pages on the WWW using your web browser. These systems will now be discussed and practical examples given for searching for information related to dentistry.

Major search engines/directory services

I will discuss these two applications concurrently since they are often mistakenly grouped together under the label of 'Search Engines' but are in fact very different tools. (It has also become increasingly common to find both types of service at one site.)

The main difference between a search engine and a directory is due to the way that they compile their data:

Search engine

Search engines use automated programs called 'robots', 'spiders' or 'crawlers' that continually visit web pages and index some, or all of the information that they find. The 'robot' can visit as many as 10 million pages per day and returns to each site on a regular basis to look for any changes. This index forms the basis for a gigantic database, the size of which varies with each search engine, but they index somewhere in the region of 30–140 million web pages. The search engine software is the program that sifts through the information in the index and rapidly finds a match to the keyword you are searching for. Examples of search engines include: AltaVista, HotBot, Excite, Infoseek and Lycos. Another search engine, Deja News, is a bit different, in that it provides a searchable archive for newsgroup articles rather than web pages.

Directory

A directory is partly put together by human researchers, who read web pages and then assign them to appropriate subject categories. Because of the human role, directories can often produce a more precise result than a search engine. On the other hand, because a directory may only cover some 800 000 sites, it may miss some of the information found by a search engine. Yahoo! is the oldest, biggest and most respected of all the directories. Other examples include LookSmart and Yell.

See Table 1 for a list of the major search engines and directory sites.

A practical example of using the AltaVista search engine

I will show how a search engine works by using AltaVista to search for some information on

---

This section of the series introduces the topic of WWW search engines and directory sites; showing you how to find information using a search engine and a directory. It also covers other ways of searching for information on the internet.

Key:

Text that is of a more technical nature

web browser

Keyword defined in the margin

password

Word(s) typed in at the keyboard

○ *Forward* ○

The name of a button (or icon) in a program

Return to home page

Hyperlink text

the internet guide for dentistry

Table 1 A list of the major search engine and directory services on the world-wide web and their features

Search Engine/ Directory	URL	Feature
AltaVista	http://www.altavista.com	One of the biggest and most powerful search engines on the WWW. It has a partnership with LookSmart who provides it with its directory listings. It also features a directory and on-line shopping.
HotBot	http://www.hotbot.com	Comprehensive; intuitive; user-friendly and fast search engine.
Excite	http://www.excite.com	Useful ability to list results by web site as well as search for documents that are similar to any given in the results. All results are given a relevancy rating.
Infoseek	http://www.infoseek.com	Runs search engine and separate directory. Results are automatically given as both web pages and as reviewed sites from its directory listings.
Lycos	http://www.lycos.com	Features 'Top 5%' which are sites that reviewers have picked as being the best on the WWW. Can also search for pictures and sounds using keyword search.
Deja News	http://www.dejanews.com	Dedicated to searching newsgroup discussions, with an archive stretching back to March 1995.
Yahoo!	http://www.yahoo.com	The most popular directory on the WWW, listing more than 800 000 web sites. It has links to all the major search engines.
LookSmart	http://www.looksmart.com	Contains a directory of 20 000 subjects covering 300 000 web sites. Also features Smart Shopper and Classified Advertisements.
Yell	http://www.yell.co.uk	Operated by BT, YELL is a guide to UK sites on the WWW. It also features an electronic version of the Yellow Pages (EYP) and an entertainment guide.

different dental topics. All search engines work by performing a free-text search on keywords to generate a list of websites that match your requested query. Your results will be different from mine since the AltaVista robot, (called 'Scooter'), visits more than 10 million web pages per day; this phenomenon can be summed up by the internet acronym 'YMMV' or 'your mileage may vary'.

1. Visit the AltaVista site by typing **www.altavista.com** in the address box of your web browser.

Fig. 1 Simple query box from Alta Vista's search engine

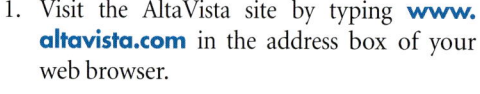

2. From the AltaVista home page you will see the simple query box which you use for everyday searches and which is common to most search engines (fig. 1). Type in the word **dentistry** and click on the button labelled ○ **Search** ○. (Use only lower case unless you want your search to be case sensitive. If you search for Dentistry, you will only get documents that include that word with just that capitalisation.)

3. The hourglass cursor will appear as the search engine software interrogates its index of millions of web pages for the word **dentistry**. After a few seconds a new page will appear with the results of your search. (You can just imagine walking into a library and shouting 'Dentistry' at the librarian!)

4. When I carried out this search, AltaVista produced a list of 482 860 matching results. AltaVista is such a powerful tool that it

often returns too much information for you to handle; it would take you months, if not years, to search through the various suggested pages. This illustrates the first principal of searching for information on the web: do not be too general in your choice of keywords.

5. You will find that the results are displayed in a ranked order (fig. 2), with the following criteria promoting a result to the top of the list:
- The query word is found at the beginning of the document, especially in the title.
- If you perform a search using two or more keywords, the query words are found close to each other in the document.
- The query word is found more than once in the document.

6. The results are listed in groups of ten; further groups can be viewed by clicking on the number bar at the bottom of the page. Each result has a title, a short introduction, the page's URL, the date it was last modified and a translation button which will translate French, German, Italian, Spanish and Portuguese text into English. Click on the title to open the page in your browser. AltaVista also lists **RealName** addresses at the top of the search results; this is aimed at making it easier for users to find company sites. For example, if you searched on the words **Hewlett Packard** you would find that the top 10 results are likely to be just web pages that mention the company or its products. However, click on the RealName link and you will be taken straight to the official Hewlett Packard home page.

7. From the results page, now carry out two new searches. Try searching on the word **dentist** and then **dental**; you will find differences in the number of results, (I obtained 291 020 and 1 241 700, respectively). This illustrates the second principal of searching: there may be many variations to a particular keyword. To overcome this you can do a third search using an asterisk (*) to act as a wildcard character. This means that searching on the word **dent*** will search for the words dentist, dentists, dentistry, dental, denture, denturism, dentate, dentition, dentifrice, etc. (Since a large proportion of web sites originate from the USA, you can also use the wildcard character to look for color and colour by using the query **colo*r**.)

8. Let us look at a more specific dental topic. Type **oral hygiene instruction** in the query box. You may be surprised to find that AltaVista returns more than 2 million matching results. Unless you state otherwise, some search engines will assume you are looking for the occurrence of ANY of your search terms. In other words, the results show any web page that contains the word

oral and/or **hygiene** and/or **instruction**. Putting a plus sign before a word tells AltaVista that each word MUST appear in the web page. For example, try searching for the following sequence of words: **+dent* +bleaching +technique**. This illustrates the third principle of searching: be clear in your mind as to whether you want to search for matches to ALL rather than ANY of your search words. Use as many different but relevant search criteria as you can in order to narrow down your search.

9. Now use quotation marks to search on the phrase **"oral hygiene instruction"**. Instead of the 2 million results that we got earlier, we now find that AltaVista returns a more manageable list of 200. This illustrates the fourth principal of searching: if you know that certain words are likely to appear together then search for the phrase rather than the individual words.

10. Another useful tip is to use a 'constraining' search. If there is a web site which is well known for providing information on a particular topic, then you can search for any other web pages that have a link to that site. For example, if there was a site called www.implants.com, that happened to be the pre-eminent site for anything to do with dental implants, you could type **link: implants.com** in the query box and find other web sites that have a link to it.

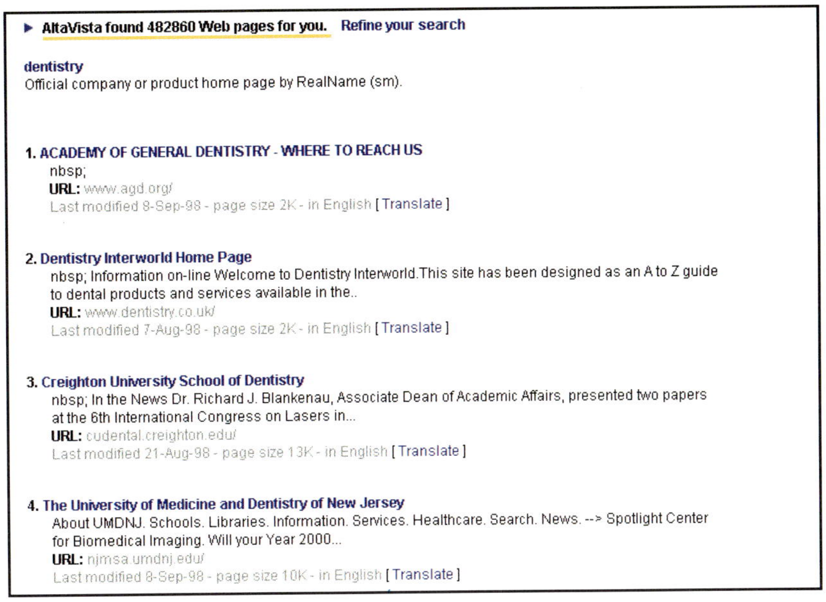

Advanced searching techniques

To the right of the simple query box on the AltaVista home page you will find a link to Advanced search. This takes you to the advanced search query box where you can extend the possibilities of your search. Here you can use Boolean logic terms such as AND, OR, NOT and NEAR. Here is an overview of how they work:
- x AND y. This forces instances of both the

▶ **AltaVista found 482860 Web pages for you.**   Refine your search

dentistry
Official company or product home page by RealName (sm).

1. **ACADEMY OF GENERAL DENTISTRY - WHERE TO REACH US**
   nbsp;
   URL: www.agd.org/
   Last modified 8-Sep-98 - page size 2K - in English [Translate]

2. **Dentistry Interworld Home Page**
   nbsp; Information on-line Welcome to Dentistry Interworld.This site has been designed as an A to Z guide to dental products and services available in the..
   URL: www.dentistry.co.uk/
   Last modified 7-Aug-98 - page size 2K - in English [Translate]

3. **Creighton University School of Dentistry**
   nbsp; In the News Dr. Richard J. Blankenau, Associate Dean of Academic Affairs, presented two papers at the 6th International Congress on Lasers in...
   URL: cudental.creighton.edu/
   Last modified 21-Aug-98 - page size 13K - in English [Translate]

4. **The University of Medicine and Dentistry of New Jersey**
   About UMDNJ. Schools. Libraries. Information. Services. Healthcare. Search. News. --> Spotlight Center for Biomedical Imaging. Will your Year 2000...
   URL: njmsa.umdnj.edu/
   Last modified 8-Sep-98 - page size 10K - in English [Translate]

Fig. 2 Results page from searching the word dentistry using the Alta Vista research engine

Newsgroups = a type of electronic discussions group

Web browser = software program that allows you to view pages on the world-wide web

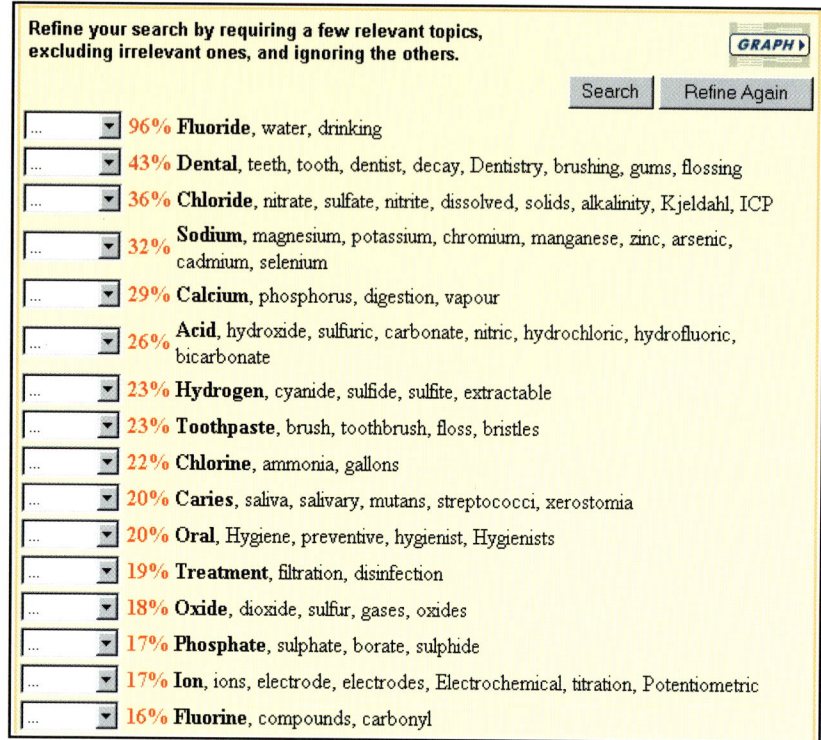

Fig. 3 The refine page from the Alta Vista search engine; the initial query word was **fluoride**

word or phrase x and y. It is similar to +x +y.

- x OR y. Will return matches of word or phrase x or y.
- x NOT y. This will reject any document with the word or phrase y in it.
- x NEAR y. This will return documents if word or phrase x is within ten words of y.

By combining these 'operators' with parentheses it is possible to form quite complex search expressions. For example, try a search on the following expression:

(root canal) AND (niti or "nickel titanium") AND technique

Note: These complex expressions only work from the Advanced Search query box; it is not necessary to type the operators in capital letters.

11. You will find a list of hyperlinks for speciality searches below the simple query box. Click on the Usenet link to search for articles that have been posted to newsgroups within the past 2 weeks. (The Deja News search engine is a better choice for searching for newsgroup articles since its archive goes back many years.)

12. An extremely useful feature of AltaVista is the ability to refine your search by adding or excluding various related search terms which are suggested by the program. Simply search for a particular topic, and when the results appear, select the ○ **Refine** ○ button which is located to the right of the query box. You will then be presented with the refine page, with a list of different groups of words related to your initial query (fig. 3). AltaVista gives each group a percentage to show you how relevant it believes the group is to your

search. You then add or exclude any of the groups listed and perform the search again.

13. From the refine page, you are also given the option of a graphical display of the related search terms; simply click on the ○ **Graph** ○ button in the top right hand corner of the topic list. This is an even more powerful tool, since from the graph page you can include/exclude individual words from each group. Try using these features to find information on **fluoride**.

14. The AltaVista web site also allows you to conduct your search by looking through various hierarchical categories that are supplied by the LookSmart directory.

Differences between AltaVista and other search engines

All search engines use the same basic technique to build a catalogue of web pages, but there are differences on how they select the data and how the query is run; that is why searching for the same word on different search engines will often produce quite different results. Always read the relevant search engine help page, since for example, HotBot will automatically search for pages containing ALL of your search words, rather than ANY of them and provides phrase searching using a menu system rather than enclosing the words in quotation marks.

A practical example of using the Yahoo! directory

A net directory is good if you want to search for some information on a general topic but do not want to be presented with thousands of results, many of which may not be relevant. The hierarchical nature of the directory and the fact that the sites have been categorised by human researchers who read the pages, means that most of the rubbish has been filtered out for you. I will show how a directory works by using Yahoo! to search on specific dental words and then use the category feature to find all the links to dental topics:

1. Visit the Yahoo! site by typing **www.yahoo.com** in the address box of your web browser.

2. From the Yahoo! home page you will see a query box. You can use this to perform a free-text search of the Yahoo! database of categories and sites.

3. The query box works in a similar way to the one in AltaVista with most of the query syntax being the same. For example, use the quotation marks to search on the phrase **"porcelain veneers"**; when I did this, Yahoo! returned no directory matches and

nine web site matches. Click on the link named web pages and you are then offered 180 web pages that match the search phrase. At the bottom of the result page are links to all of the other major search engines; clicking on one of these links will automatically carry out the same search using that search engine. In this case AltaVista produced 1137 results, HotBot 309, and Infoseek 344.

4. If no matching Yahoo! categories or sites are found, Yahoo! will automatically perform a web-wide, full-text document search using the Intkomi search engine.

5. The strength of Yahoo! is the list of categories, which is located just below the query box on the Yahoo! home page (fig. 4). From the 14 main headings, it is possible to 'drill down' through the categories into more narrowly focused headings. Click on the Health category and you will be taken to a page of sub-topics, one of which is labelled Dentistry.

6. Click on the Dentistry link and a new page will appear which lists all of the dental sub-categories as well as some of the dental sites. Each site listing is a hyperlink to that site and beside the link is a brief explanation of what can be found at that location.

Fig. 4 The Yahoo! directory home page. Dentistry can be found by clicking on the Health hyperlink

Regional editions of search engines and directories

The major search engines index web pages from all around the world. Recently, however, regional editions have appeared to serve those people living in particular countries or regions. The main reason for their existence is to enable better targeting for the advertisers who pay to appear on the search engine or directory. Some editions just provide a different interface, (for example, the commands could appear in the French language), while others give true local content. Some regional sites are physically located outside the United States, (called a 'mirror' site), and this can improve responsiveness when the American sites are busy. However, the mirror sites do not normally hold as big an index of data as main sites, nor is the information as up-to-date.

Yahoo UK & Ireland can be found at **http://www.yahoo.co.uk** and it offers local content such as news and internet events. When you carry out a search you can specify for it to search all sites, (the default selection), or just those sites in UK and Ireland.

Search engine software for finding information within a web site eg Excite Web Server (EWS)

There are countless software programs available that authors can add to their own web sites to enable users to search for information within the site. An example of this is the free search engine called Excite Web Server

(EWS) and is available to download from the Excite site. You will find that most large web sites include some sort of local search facility.

Software that enables you to query a database of information by the internet eg MEDLINE

Doctors and dentists have been using MEDLINE in libraries for years to help with citation searches of published biomedical literature. You can now carry out the same searches from the comfort of your home by visiting one of the many sites on the internet that offer this free service.

The best site to use is PubMed[5] since it is run by the National Library of Medicine who provides the main source of data to the other web sites. PubMed has the most up-to-date information and unlike other sites, it does not require you to register in order to use the service. The PubMed database is drawn primarily from MEDLINE and PREMEDLINE (which gives basic information prior to the full record being added to MEDLINE). It also has links to full-text journals at Web sites of participating publishers.

The MEDLINE data is stored on computer as a large database and the search button on the web page enables you to send a query to this database. Once it has searched the database using your keywords, it creates a new web page that lists the results. In other words, the MEDLINE data itself is not stored as web pages and because of this you cannot use an ordinary search engine such as AltaVista to carry out a MEDLINE search. However, PubMed does provide a WWW Citation Matcher service, which allows publishers to match up their own citations to PubMed entries. This permits publishers to link from references in their published articles directly to entries in PubMed.

A practical example of using PubMed

Let us imagine that you remember seeing an article in the *British Dental Journal (BDJ)* by Peter Briggs on the subject of tooth wear and you want to find out more about it.

You can search on one or more terms (eg **tooth wear**) by typing them into the query box and pressing the ○ **Search** ○ button (265 results). You could also just search on the author's name using the format of the surname plus initials (no punctuation), eg **briggs p** (86 results).

Not all of the 86 results were authored by P Briggs; why is that? PubMed automatically truncates on the author's name to account for varying initials and designations such as Jr. Therefore, if you want to be more specific about the author use double quotes around the surname and first initial and add the author search field tag [au], e.g., **"briggs pf" [au]** This will turn off the automatic truncation and retrieve results based on only the single first initial.

If you now carry out a query using the combined search terms **tooth wear briggs p** you will retrieve a useful list of 8 references, 2 of which are from the BDJ. (To show only those articles that were published in the BDJ you could have added either the full journal title, ie **british dental journal** or its MEDLINE abbreviation **Br Dent J** to your search).

From the search results page you can view the abstract from a single article by clicking on the author's name. If you wish to see all of the abstracts from your results, (where available), click on the ○ **Display** ○ button. If just want to see certain of the abstracts, click the box to the left of the author's name for those articles that interest you and then click the ○ **Display** ○ button.

MeSH

MeSH is a vocabulary of medical and scientific terms assigned to documents in PubMed by a team of experts. In many cases, these terms can be used to search PubMed more efficiently than just using simple text words. From the PubMed home page, click on the link to the MeSH Browser. Type in the term dentistry and you will see all the relevant dental MeSH terms displayed in a hierarchical structure

eg Tooth Preparation
    Dental Cavity Preparation
    Root Canal Preparation
    Tooth Preparation, Prosthodontic

Continuing with this example search, you will find that one of the two citations from the result page is a letter commenting about a previous article in the *BDJ* while the other is the article you wanted to find (*The clinical evaluation of the "Dahl Principle"* by Briggs P F, Bishop K and Djemal S). You now have the option of:

· Reading the abstract of the article
· Printing the abstract using your web browser ○ **Print** ○ button. (Copyright limits this for personal use only.)
· Adding the result page to your browser Favorites or Bookmark list.
· Saving the text as a file using the ○ **Save** ○ button at the bottom of the result page.
· Ordering the full document from a local library with whom you have set up a Loansome Doc agreement. Use the ○ **Order** ○ button at the bottom of the result page.
· Searching for other articles in MEDLINE that closely relate to the selected article by clicking on the ○ **Related Articles** ○ button (107 results).

Gateways to evaluated, subject-orientated information

Directories of evaluated information are becoming an important search tool for medical and dental professionals. There is no shortage in the quantity of information on the internet, but sometimes it can be difficult to find quality material. As we have seen with AltaVista and Yahoo!, a search will normally offer high returns in regards to quantity, but the results may be questionable in terms of the relevance and quality of the information provided. It is the age-old problem of sorting the wheat from the chaff.

In an article in the *BMJ*, a survey on WWW advice on how to deal with a childhood fever highlighted the questionable accuracy of medical information on the internet.[5] The authors looked at 41 web pages relating to home management of feverish children and found that only four web pages adhered closely to published guidelines.

Most information on the internet has not undergone any form of peer review and because of the ease of publishing a personal web page, there has been an explosion in the field of home publishing, sometimes referred to as 'vanity publishing'. It is often unclear whether the information is targeted at the lay public, the professional or the academic. Additional problems are that the information may either be out-of-date or that the referenced web page no longer exists.

Some popular search engine sites do try to review the best web resources, for example the Top 5% catalogue at the Lycos site,[6] but the criteria used are often very informal, such as 'how cool is the site?'. Do not assume that any dental web site that proudly displays stars or badges awarded by such sites has necessarily had any form of critical evaluation.

In recognition of the need for access points to quality, subject-orientated information, a number of gateway sites have appeared which attempt to guide its users towards evaluated resources. In the area of health and medicine, the main well-established internet guides are HealthAtoZ,[7] Healthfinder,[8] HealthWeb,[9] Medical Matrix,[10] and Six Senses.[11] However, these sites have a very American slant to them; I would highly recommend the excellent UK gateway site, OMNI.[4]

OMNI: organising medical networked information — a practical example of using a gateway to evaluated medical/dental information

OMNI is the UK's main gateway to high quality biomedical information on the internet. It aims to filter, evaluate, describe, index and classify UK biomedical resources comprehensively, as well as a number of select worldwide sites. At the start of October 1998, OMNI had catalogued more than 3700 resources. All records are searchable or you can browse the 55 subject categories. There is an additional search engine called Harvester which was experimental at the time of writing. Harvester is composed of resources recommended by authoritative lists of internet resources compiled by subject experts and therefore covers thousands more sites. OMNI also has a facility which allows you to browse the DERWeb Dental Image Library as well as supporting access to MEDLINE references directly through hyperlinks.

Hyperlinks from individual web pages, newsgroups articles and e-mail messages

The last, but most obvious way of finding information on a topic is by following related hyperlinks from web pages, newsgroup articles or e-mail messages. A good example of links from web pages is the Electronic Telegraph.[12] If you find an interesting news article and want to have more background information about the subject, you can either click on one of the internal hyperlinks to other related Telegraph stories or on one of the external links to other web sites. For example, a story about trouble in Iraq would have internal links to other related articles in the newspaper as well as external links to the UN Security Council, the Iraq-Arab Net and to the CNN web site.

Most web sites contain a page which consists solely of links to other related sites. For example, you will find DERWeb's links to other interesting dental sites at:

http://www.derweb.ac.uk/mid6.html

Which search tool should I use?

Search directories such as Yahoo! are a good place to start a search when your topic is broad. For example, you can use a directory to find general information about orthodontics by 'drilling' through the categories until you eventually find a shortlist of reviewed orthodontic sites. (If you used a search engine, such as AltaVista, to search on the word **orthodontic** you would be overwhelmed by the thousands of matching web pages). For a complete listing of every single web page that matched a well thought out list of keywords, then use a search engine. To view evaluated biomedical web resources use a medical gateway site (such as OMNI) and for very specific biomedical information use MEDLINE as your search tool of choice.

1. AltaVista:
   http://www.altavista.com/
2. Yahoo!:
   www.yahoo.com/
3. PubMed:
   http://www4.ncbi.nlm.nih.gov/PubMed/
4. OMNI:
   http://www.omni.ac.uk/
5. Impicciatore P, Pandolfini C, Casella N, Bonati M. Reliability of health information for the public on the World Wide Web: systematic survey of advice on managing fever in children at home. *BMJ* 1997; **314**: 1875-1879.
6. Lycos Top 5%:
   http://point.lycos.com/
7. HealthAtoZ:
   http://www.healthatoz.com/
8. Healthfinder:
   http://www.healthfinder.gov/
9. HealthWeb:
   http://healthweb.org/
10. Medical Matrix:
    http://www.medmatrix.org/index.asp
11. Six Senses:
    http://www.sixsenses.com/
12. The Electronic Telegraph:
    http://www.telegraph.co.uk/

Dental resources on the internet

In the 3 years that I have been using the internet, the amount of dental information on the internet has increased dramatically. It is true to say that the majority of it still comes from the USA, but the quantity and quality from UK sites is now starting to flourish.

In this part of the series I have:

- Listed web sites which I think dental practitioners would find interesting and useful, (additional good sites can be found in the references to each part of the series).
- Divided the internet resources into dental, business and computing.
- Categorised the sites according to the type of resources available, see Table 1. (Remember that some of the resources may be subject to copyright.)
- Selected web sites which use the powerful features of the internet ie the ease of searching through masses of information; multimedia elements; the currency of information; the fact that the site is regularly updated; and the information is neither easily nor freely available using any other medium.

It is impossible in just a few pages to list every dental resource on the internet; if you want a more comprehensive list then either look at the dental link sites mentioned later on in this section or peruse the recently published book from Quintessence, *The global village of dentistry*.[1] The information on the internet is changing at such a massive pace that any list can really only be thought of as a stepping stone onto your own unique exploration.

DENTAL RESOURCES ON THE WWW

American Academy of Pediatric Dentistry (USA)
http://aapd.org/

→ I A ○

The AAPD's 3700 members are dedicated to improving the oral health of infants, children, adolescents, and patients with special healthcare needs. The site lists answers to questions that parents commonly ask about their children's teeth. The 20+ press releases make interesting reading. It has extensive links to general 'parenting sites' on the internet.

American Academy of Periodontology (USA)
http://www.perio.org/

⌘ → A P

Some of the content of this site is limited to its 7000 members, but you can access the list of nearly 30 position papers, statements and parameters (requiring *Acrobat Reader*). There are also good links to other periodontology web sites.

American Dental Association (USA)
http://www.ada.org/

⌘ I A

This is an extensive site but much of the information is not relevant to the UK; some of the content is for ADA members only. It is worth visiting for the abstracts from the current issue of JADA, articles from the ADA News, the FAQ section for patients and the section on 'research and clinical issues'.

Australian Society of Orthodontics (AU)
http://www.aso.org.au/

→ I A ○

This site has a very good section on information for orthodontic patients and includes some good images; it also contains links to other orthodontic sites.

Bad Breath Research (I)
http://www.tau.ac.il/~melros/welcome.html

→ I U

Tel Aviv University have produced a page on 'questions and answers' to halitosis, information about the Society for Breath Odour Research and links to related sites.

British Dental Association (UK)
http://www.bda-dentistry.org.uk/index.html

A

Contains details about members services, a selection of fact files and the full text of press releases going back to January 1996.

British Dental Journal (UK)
http://www.stockton-press.co.uk/bdj/

⌘

At the moment you can only view the contents (but not abstracts or the full text) of the BDJ, going back to January 1997. There are plans to make the site far more comprehensive.

British Dental Trade Association (UK)
http://www.bdta-dentistry.org.uk/

→ T

The BDTA site is useful for tracking down a dental company, product or service in the UK.

Clinical Research Associates (USA)
http://www.cranews.com/

⌘ A M R

CRA was founded in 1976 to evaluate dental

This part of the series guides you to some of the best dental resources on the internet. There are also suggestions for the location of other useful internet resources. Advice is given on what additional software is required to make the most of these resources.

Key:

password
Word(s) typed in at the keyboard

web browser
Keyword defined in the margin

Return to home page

Hyperlink text

the internet guide for dentistry

Table 1 A guide to the symbols used to categorise the information found on the web sites listed in this part of the series.

Symbol	Topic
$	Business management
💾	Computer software
📁	Case studies
☎	Forum, discussions
✎	Graphics/photographs
📖	Journal/magazine
➡	Links to other sites
I	Patient's information
A	Associations
E	Endodontics
I	Implants
M	Dental materials
O	Orthodontics/paedodontics
P	Periodontics
R	Restorative/prosthetics
S	Oral and maxillofacial surgery
T	Dental trade
U	University/college/school

Screenshot of the BDA website

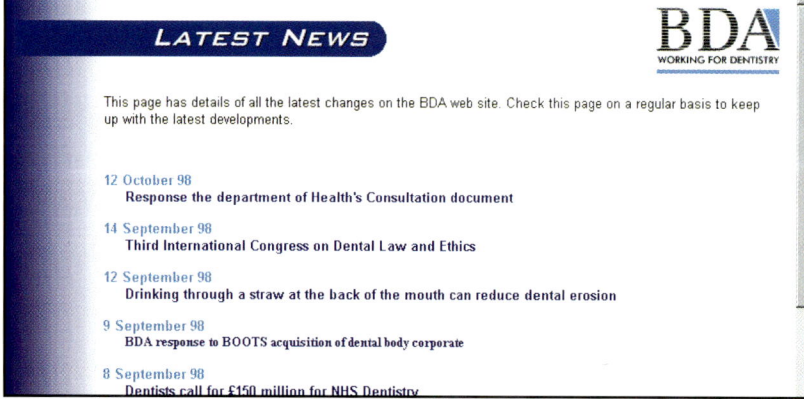

products to confirm their efficacy and clinical usefulness. Their motto is 'CRA tests so you don't have to'. Some useful information on intra-oral cameras and dentine adhesives. To view some of the data, you will require a copy of *Adobe Acrobat Reader*.

Creighton University (USA)
http://cudental.creighton.edu/
💾 ✎ O U

This site is a lovely example of how to promote a dental school. Located in Omaha, Nebraska, Creighton is a Catholic university operated by the Jesuits and has trained dentists since1906. It has an interesting page on the history of dentistry and a large catalogue of paedodontic and orthodontic images as well as a sample of a multimedia interactive program for paediatric dentistry.

Dental Bytes MagEzine (USA)
http://www.sybor.com/dentalbytes/
➡ T

This dental 'webzine' has been running since 1996 and has an informal and fun style. There is an archive of past feature articles and it contains links to other award-winning dental sites.

Dental Implant Summaries
http://www.dentalsummaries.com/
📖 I

This is the first dental implant journal available on-line and provides the practical solution for anyone wanting to keep up-to-date in implant dentistry. Here you can access the fully indexed publication back to 1993. Relevant articles in one particular field are grouped together in 33 subject categories. There are six issues published each year and they summarise topical research selected from more than 40 leading journals. Full access to all the articles is limited to paid subscribers. You can reproduce the original pages or a whole journal as it appears in the print version using *Adobe Acrobat Reader*.

Dental Shopping Mall (USA)
http://www.dentalshop.com/
➡ T

At this site you can request product information directly from dental manufacturers through the 'product search' feature. Very USA-orientated but the site is improving all the time; one to watch out for.

Dental Success On-line Magazine (USA)
http://www.dentalsuccess.com/
$ 📖

This site consists of 'articles and insights by the dental industry's leaders and opinion-makers'. The archives go back as far as May 1996; some of the articles on practice management would be of interest to private practitioners in the UK.

Dentagraphics (UK)
http://www.dentagraphics.co.uk/
✎ E I O R

This site describes how you can use a unique system of keystroke vectoring to produce sequential rows of dental images. John McCormack is to be congratulated for creating the Dentagraphic fonts which may be downloaded free (provided the conditions of copyright and use are fully met). There are eight different True Type fonts, (for either Windows PC or Mac), which can produce nearly 500 different pictures of teeth, roots, crowns, attachments, implants, endodontically filled teeth and orthodontic images. All the images are high quality vector graphics so that they can be enlarged without any loss of quality. Ideal for producing presentations or practice leaflets.

Dentanet (UK)
http://www.dentanet.org.uk/
💾 ☎ 📖 T ➡

It is now 2 years since the DPB, in collaboration with the Department of Health, set up the Dentanet site. Include yourself in the database of more than 3800 UK dentists in the 'find a dentist' section. Other facilities include a discussion forum, home page service, Product Finder, DPB statistics and quite a few trade areas such as the BDTA and *Dental Practice Magazine*. There are also currently ten CAL programs available for downloading.

DERWeb (UK)
http://www.derweb.ac.uk/
💾 📁 ☎ 📖 ✎ ➡ A T U

This site has been described as the 'jewel in the crown' of networked education and has recently gone through the transition of being funded to becoming self-funding. When I printed out the contents page from the web site it produced ten sheets of A4 paper! Main attractions are the image library, the web conferencing and live chat, the list of dental contact addresses, the DERWeb bookshop, the Schottlander quiz, the CAL pages, the educational case studies and the hundreds of links to UK and world-wide dental associations and societies.

Lund University, Faculty of Odontology (SE)
http://www.odont.lu.se/depart.html
💾 📁 ✎ P R U

Explore the fascinating periodontology and cariology departments at the Faculty of Odontology. Visit the Cariology WWW Museum and find out about the interactive Cariogram PC-program. Find out who has the best/worst teeth in the world from the WHO Oral Health Country/Area Profile Programme. Improve your knowledge at the Virtual Periodontology pages and you are invited to e-mail any questions you have about periodontology directly to the Department of Periodontology.

Mining Co.; Dentistry (USA)
http://dentistry.miningco.com/
☎ ➡

The Mining Co. runs a comprehensive network of web sites that cover more than 500 topics. The volunteer author for the dentistry site is Dr John Brooke and he has put together a nice site with a series of short articles, a chat room/bulletin board and links to other useful dental sites.

Medical University of South Carolina; College of Dental Medicine (USA)
http://www2.musc.edu/dentistry/top40/Headers/main6.htm
U

The 'TOP 40 Prescribed Drugs' provides detailed information about the most frequently prescribed medications in the USA in 1998 and their significance to the dental practitioner. The detailed database can be accessed by generic name, trade name, manufacturer, frequency prescribed, or the class/type of medication.

OMNI: Organising Medical Networked Information (UK)
http://www.omni.ac.uk/
➡

There was a brief description of OMNI in an earlier part of the series. OMNI is recognised as the most comprehensive source of evaluated biomedical resources for UK healthcare professionals. At the moment it has just under 100 dental resources listed.

Practical Endodontics (USA)
http://www.endomagic.com/
📖 ➡ E

This site contains sample articles from *Practical Endodontics* going back to 1991 plus an article entitled 'Twelve secrets of the painless 30-minute root canal'. It also has links to other endodontic sites, such as the 'Automated Endo' web site.

Procter and Gamble's Dental ResourceNet™ (USA)
http://www.dentalcare.com
🗁 ✎ 📖 I T U

A glossy site with good patient education resources and topical articles on current issues in dentistry. You need to register (free) to gain access to some parts of the site. There is a virtual classroom which has been designed in conjunction with the University of Louisville School of Dentistry. On successful completion of a course, the participant is awarded 2-4 hour CE credits which are recognised by the American Dental Association.

PubMed (USA)

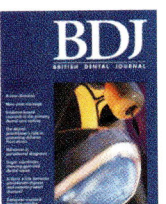

British Dental Journal

• Home
• Journals
• What's new?
• Subscribe
• Information

• Contents
• Editor
• Instructions to authors
• Subscribe
• Audience
• Related journals
• Scope
• Indexed in
• Society
• Contacts
• News
• Sample copy

ISSN 0007-0610
1998 Volumes 184/185
Published twice a month

The British Dental Journal is the official journal of the British Dental Association and contains refereed scientific papers and articles on the latest advances and findings in dentistry.

STOCKTON

Screenshot of the BDJ website

http://www4.ncbi.nlm.nih.gov/PubMed/
➡

This is probably one of the most useful sites on the WWW for healthcare professionals. The National Library of Medicine's search service gives you free access to MEDLINE with nine million citations dating back to 1966. Many sites on the internet offer free access to MEDLINE but this is the definitive site to access it from.

Temple University School of Dentistry; Department of Dental Informatics (USA)
http://www.temple.edu/dentistry/di/
📖 ➡ U

They say that America leads the way and we follow; if this is the case then this site will give you an idea of how dental informatics may be taught in the UK in the future. The university has spent more than $4 million since 1989 on computer hardware, software and support to ensure that their students are prepared for the twenty-first century.

TMJ Tutorial (USA)
http://www.rad.washington.edu/Anatomy/TMJ/TMJ.html

Screenshot of the Dental Implant Summaries website

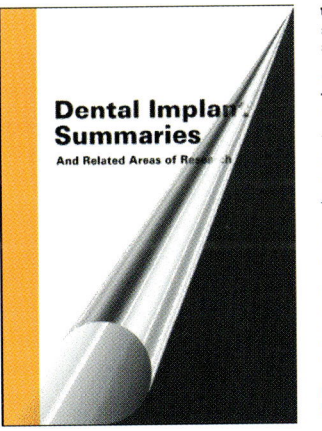

Dental Implant Summaries

Welcome to Dental Implant Summaries - the first dental implant journal available on-line. Here you can enjoy the interactive benefits of the WWW and access the publication back to 1993.

Summaries with a difference! - Far more comprehensive than abstracts, with more detail in the 'introduction' and 'discussion' to place each topic in its appropriate context.

Six Issues published each year - summarising what we believe to be the most interesting and topical research selected from more than | 40 leading journals |

33 Subject Categories - There are now more than 500 articles in this database, all conveniently arranged into 33 subject categories for simple research - | Subject Index |

Diagrams - From 1996 onwards 'diagrams' have been included in the summaries to give even greater clarity to the subject matter. To access the summaries and diagrams all you have to do is download the Adobe Acrobat PDF files accompanying each page, and then print off your own copy which looks just like the original.

- If you are already familiar with PDF files and also have copy of Acrobat Reader, then activate the PDF logo icon on the left and view a sample article in this versatile format.

[Subject Index | Latest issue | Select Issue | Feedback]   [Home | Subscription | Information | Copyright]

the internet guide for dentistry

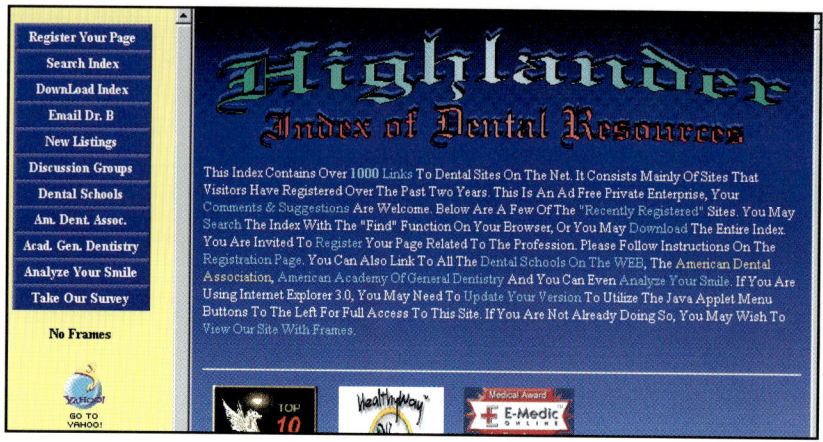

Screenshot of the Highlander website

⊟ ✎ U

Produced by the University of Washington Department of Radiology, this short tutorial shows images of TMJ anatomy, arthrography, computed tomography and magnetic resonance imaging. You can also download *Quicktime* movies showing normal TMJ function and then showing anterior displacement with and without reduction.

University of Birmingham, School of Dentistry (UK)

http://www.bham.ac.uk/dentistry/

⊟ 🗁 ✎ ➡ U R

Follow the <u>Undergraduate Dentistry</u> hyperlink to a list of UK web sites that are particularly useful for dental undergraduates. From the home page, click on the <u>Hot Topics</u> hyperlink to take you to a section that catalogues the large amount of Computer Aided Learning (CAL) material that Birmingham have developed. The <u>MedWeb</u> link leads you to 50 MCQs on immediate and complete dentures. There is also a working demo of the knowledge-based system (RaPiD) for designing removable partial dentures.

University of Minnesota; Minnesota Dental Research Center for Biomaterials and Biomechanics (USA)

http://web.dent.umn.edu/index.htm

M U

This site shows various research projects, such as the 3D simulation of jaw movements and the 3D tooth profiling system developed for the accurate measurement of tooth wear. A wide range of pictures and videos are used to illustrate the various projects.

University of North Carolina; Dept. of Oral and Maxillofacial Radiology (USA)

http://www.dent.unc.edu/depts/diag/newslett.htmSU

🗁 S U

This page shows the *Teledentistry Network Newsletters;* a collection of case studies with relevant radiographs and tomograms to enable you to work out a differential diagnosis.

UMDS Dental Materials Conspectus (UK)

http://www.umds.ac.uk/dental/dms/conspect.html

MTU

Dr David Brown has put together a list of more than 100 UK suppliers of dental materials, and more than 2000 materials with a brief description of what the materials are (or were).

DENTAL LINK SITES ON THE WWW →

Highlander list of dental resources (USA)

http://www.mindspring.com/~cmcleod/highlander.html

This site boasts links to more than 1000 dental resources. They even provide a selection of background music to listen to while you browse the site! Download the entire index as a zip file to browse it off-line. The ability to view just the recent additions is a useful feature.

Dental Related Internet Resources (USA)

http://www.dental-resources.com/

If you are just looking for links to other dental sites then this should be your first stop. You can either search the index or use the extensive directory. The information is updated almost daily and the site receives more than 46000 hits per month.

Internet Dentistry Resources (USA)

http://indy.radiology.uiowa.edu/Beyond/Dentistry/sites.html

This site contains a reasonably well updated list of dental links.

The Virtual Dental Center — Martindale's Health Science Guide (USA)

http://www-sci.lib.uci.edu/~martindale/Dental.html

The Martindale Health Service Guide is a gigantic site currently containing thousands of teaching files, medical cases, multimedia courses/tutorials, databases and movies. The Dental Center is part of this site and has well organised links.

DENTAL MAILING LISTS ☎

The GDP-UK Mailing List was started in 1997 by Tony Jacobs, a GDP from Manchester and now has more than 100 participating members from the UK. The main topics for discussion have centred on dental computing, dental materials/products, dental politics and post-graduate education. To join the list, send an e-mail to tony@jacobs.net

The Internet Dental Forum is a 'daily electronic exchange of dental tips from around the world'. Although there is a definite American bias, this mailing list has an international membership of more than 1200 dentists, technicians and hygienists. The only problem with this discussion forum is its popularity; the daily digest can often exceed 80 messages. To join the list, send an e-mail to David Doddell at david@stat.com

The University of Iowa College of

Dentistry

http://indy.radiology.uiowa.edu/
Beyond/Dentistry/leslie.html

This page contains a list of more than 30 electronic discussion groups in dentistry and includes a brief description of who each group is aimed at.

Liszt

http://www.liszt.com/

This site enables you to search through a database of more than 90000 mailing lists available on the internet.

USEFUL BUSINESS SITES ON THE WWW $

a2btravel (UK)

http://www.a2btravel.com/

If you are travelling across the country for a conference or course, (or even a holiday!), then this site will be of great assistance. It covers everything; on-line timetables for trains and ferries, on-line booking of flights and accommodation, (with a price comparison and availability checker), live flight arrivals into the UK, tube/subway maps for the major cities and much, much more.

CCTA Government Information Service (UK)

http://www.open.gov.uk/

The Government Information Service home page is not very intuitive when it comes to navigating around this vast site. Use the Functional Index or search facility to direct you to the Inland Revenue and Data Protection Register pages.

Contributions Agency (UK)

http://www.dss.gov.uk/ca/

Check out the latest National Insurance information for employers and employees.

MoneyWorld UK

http://www.moneyworld.co.uk/

This is one of the most comprehensive personal finance web sites for the UK; guides to personal finance topics, news on the latest budget and share prices, a glossary of financial terms, financial contact list, and rates and performance of unit trusts, mortgages and savings.

USEFUL COMPUTING SITES ON THE WWW

PC Webopaedia (USA)

http://www.pcwebopaedia.com/

The leading on-line encyclopaedia and search engine dedicated to computer technology. It features a 'term of the day' and has a listing of the top 15 terms requested.

ZD Net (USA)

http://www.zdnet.com/

ZD Net publish over a dozen computing magazines in the USA and has links to all of them; a treasure-trove of news, information, tips, reviews and tutorials for the computer enthusiast.

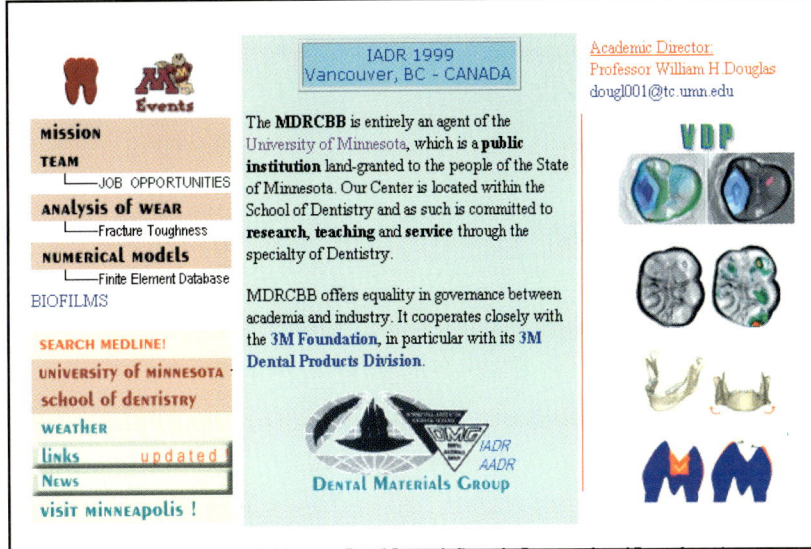

Intel's Connected PC (USA)

http://www.connectedpc.com/

This site contains entry level information on what you can do with your PC once you are connected to the internet with simple practical applications. Examples include exploring virtual reality, and making an internet phone call.

Stroud's Consummate Internet Apps List: (USA)

http://cws.internet.com/

This is the ultimate one-stop download site for the latest internet application software. The site provides comprehensive file listings, ratings, and reviews for the best Windows 95/98/NT and Windows 3.x internet software.

ADDITIONAL SOFTWARE REQUIRED TO MAKE THE MOST OF THE LISTED RESOURCES

Adobe Acrobat Reader

http://www.adobe.com/

Freely available, *Adobe Acrobat Reader* allows you to view documents created in this format; Acrobat files have the extension .pdf. When printed, pages are easier to read than standard WWW pages because they retain the original design and layout.

Apple Quicktime

http://www.apple.com/quicktime/

Quicktime 3 allows you to play Quicktime video and sound files downloaded from the internet. *Quicktime VR* allows you to view virtual reality files such as those found at **http://www.nidr. nih.gov/iyf/vanatomy.html** of the head and skull.

RealNetworks

http://www.real.com/

RealPlayer (version 5 or the latest G2) is used for playing live or recorded audio/video over the WWW. This is made possible by playing the multimedia file while it is still downloading to your computer; a process known as streaming.

Screenshot of the Minnesota Dental Research Center for Biomaterials and Biomechanics website (Reprinted with the permission of Dr W H Douglas and Dr T Korioth)

1 Spallek H., Spallek G. *The global village of dentistry*. London: Quintessence, 1997.

Efficient use of the internet

People often complain about how slow the internet is and how it can be a time-waster. However, there are many things within your control that can make your time on-line more productive and enjoyable. These include the following:

1. Organising and planning your on-line sessions.
2. Getting the best from your existing hardware and software.
3. Using additional hardware and software.

In this article, the tips for using the WWW will work with most versions of web browser. In some examples I have described the specific menu bar commands for different versions of the most popular web browsers (Microsoft Internet Explorer version 3 and 4 (MSIE3 and MSIE4) and Netscape Navigator version 3 and Netscape Communicator (NN3 and NN4)).

Organising and planning your on-line sessions

- Plan your internet session before going on-line, just as you would plan what you were going to say before making an important telephone call. If you are going to use the WWW, then have a clear idea of what you are going to use it for.
- Get into the routine of checking your e-mail almost every day. A recent survey on the use of the internet by dentists in the UK showed that 85% considered e-mail to be their main use of the internet and 76% used the internet every day.[1]
- Do not join more newsgroups than you can easily manage; it is very easy to develop a severe sense of 'information overload'. Unsubscribe from groups when you know you will be too busy to read the articles or you are going on holiday. Unless you think you will be interested in reading the majority of the articles from a newsgroup, set up your software to retrieve only the article headers, rather than the whole body of the text.
- If you want to keep up-to-date with newsgroup discussions, run your newsgroup program every time you run your e-mail program.
- Plan ahead by creating a simple text file using a text editor such as Windows or *Word Pad* or *Notepad* and place a shortcut to the file on your desktop. Call the file something like *web sites to visit* and type in or copy/paste interesting web page addresses which you come across in newspapers, magazines and television or while you are reading your e-mail or newsgroup articles off-line. Once you are on-line you can copy and paste the URLs into the browser without having to waste time typing them in. You can also use the text file to make notes about searches you wish to carry out for the next time you are on-line. (If you use NN, a neater option would be to create new bookmarks off-line.)

- Before you go on-line, (provided your software supports it), open the various internet application programs which you are planning to use during that session.

- With most web browsers, it is possible to configure which page loads when you open the browser (the 'opening' or 'start' page) and which page loads when you click the ○ **Home** ○ or ○ **Search** ○ button. A lot of people never bother to change the default settings that come with the browser, yet this is very easy to do and can make your browsing a lot more efficient.

In NN3 you can set up the browser to either start with a blank page or else the home page. You can specify the home page to either be a page on the internet (eg the home page of your favourite web site) or a page located on your hard disk. The advantage of having it set to a page on your hard disk is that the page will load much faster than any page on the WWW and it gives you an opportunity to create your own custom made page with links to your favourite web sites and search engines.

The easiest way to create a custom page in NN3 is to use the fact that NN stores its bookmarks as an HTML document called bookmark.htm and stores this file in the Netscape directory; edit your bookmark file using *Notepad* and save it under a different name eg *mylinks.htm*. Then go to **Options**, **General Preferences** and select the tab labelled **Appearance**; in the section '**Browser Starts With:**', select the ○ **Home Page Location** ○ radio button and type in the file name and its location on your hard disk; see figure 1. (Unlike NN, where all your bookmarks are stored in a single HTML file, MSIE instead creates a shortcut for each favorite that you have. If you use MSIE, you can use a utility called *Bookmark Importer*[2]

This section explains how to make most of your time on-line by providing tips on how to reduce the cost of using the internet. Suggestions are given on how to get the best from your existing hardware and software. There are also recommendations for some additional hardware and software.

Key:

Text that is of a more technical nature

password

Word(s) typed in at the keyboard

web browser

Keyword defined in the margin

readme.txt

The name of a file (or document)

○ **Forward** ○

The name of a button (or icon) in a program

File, Open

Step-by-step instructions using the menu bar

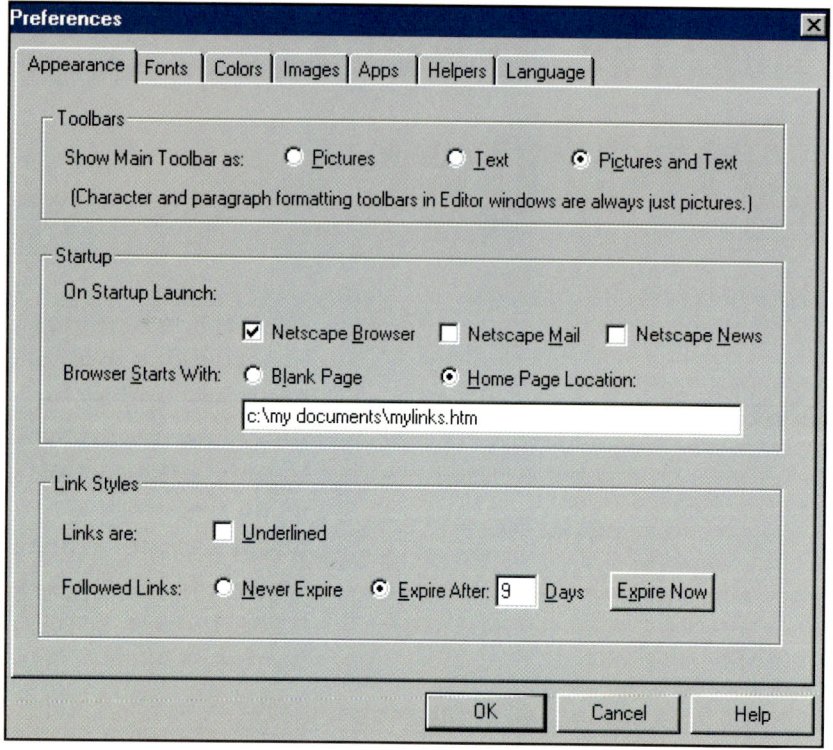

Fig 1 Screenshot showing the window in Netscape Navigator version 3 where one configures the location of the Start Page. In this case it is located on the hard disk and the page has been named *mylinks.htm* **(Reprinted with the permission of Netscape Communications Corporation)**

(see later in this article) to export your Favorites directory to an HTML document and use this as your home page.)

I have created a customised start page that you can download from my web site.[3] It contains links to useful dental, business, computing and search web sites. You can save this page on your hard disk and use it as your opening page. If you know a little bit about HTML you will easily be able to edit it using *Notepad*.

In NN4 you can choose the opening page to be a blank page, the home page or the last page you visited in your previous session online. Select **Edit**, **Preferences** and click on **Navigator**. You can add items to the Personal Toolbar by dragging the location icon for the currently viewed page (found next to the address bar) onto the Personal Toolbar.

In MSIE3 you can select which page to use for your start page, search page and the hyperlinks on the Links toolbar. If you wanted to change

the setting for the start page to a page stored on your hard disk, click on **View**, **Options** and then select the **Navigation** tab. In the section called **Customize** click on the drop-down box and select **Start Page**. Now type in the directory path for your start page, or first open the start page and then select **Use Current**. The same drop-down box can also be used to customise the search page and the five 'Quick Links' in just the same way (see figure 2). You could use these for creating quick links to your most frequently visited web sites (eg DERWeb or MEDLINE) and the appropriately labelled buttons would then always appear in the Links toolbar.

In MSIE4 you have the choice of having the opening page either as a blank page, the default page (which is normally the Microsoft home page or the home page of your particular internet provider) or use the current page you are viewing. Select **View**, **Internet Options** and then click on the **General** tab. You can easily add items to the Links toolbar by dragging the icon for the page from your Address bar to the Links bar.

- Write all your e-mail messages and newsgroup articles before you go on-line.
- If you intend sending someone an e-mail attachment, make sure the file is as small as possible by compressing it with a program such as *WinZip*.[4]
- Decide how long you ideally want to spend on-line before you make your connection. It is a well known phenomenon that 10 minutes spent on the computer equates to 60 minutes in real time! Use an egg timer to remind you when your time is up. To prevent strain injuries, take a 5-minute break every half an hour.
- Picking the best time of the day to go on-line can make a huge difference in how quickly your web browser will download pages from the WWW. It is a combination of how busy the connections are across the Atlantic and how many subscribers to your internet provider have also decided to log on at the same time. My own provider (Demon) have produced some interesting data that showed that, as far as local system usage was concerned, the peak time band was from 6 pm to 11 pm during the week. There was normally a trough at 5 pm as people made their way home from work, and every Thursday evening at 9.30 pm there was a dip when *Men Behaving Badly* was being broadcast on television! Keep in mind that US Pacific Time is 8 hours behind GMT — so most people on the West Coast are asleep when it is mid-morning in the UK. Because of these usage patterns, I normally retrieve and send my e-mail messages and newsgroup articles every weekday evening, (which normally takes less than 3 minutes), and browse the

Table 1 British Telecom's pricing structure; pence per minute for a local call (inclusive of VAT). There is a minimum charge of 5p for any call. Prices are correct as of 01.11.98

British Telecom	Pence per minute for a local call	Cost of a one-hour call
Daytime Mon to Fri 8 am – 6 pm	3.95p	£2.37
Evenings and night-time Mon to Fri before 8 am and after 6 pm	1.49p	90p
Weekend Midnight Fri - Midnight Sun	1p	60p

WWW and download programs during weekend mornings. It is a bit like getting to know what are the best times to travel to avoid the traffic; you may still need to travel during rush hours but don't be surprised if your journey takes longer.

- Check with your telephone company how you are charged for local rate calls; keep an eye on British Telecom's (BT's) charges by visiting their web site.[5] Table 1 shows details of BT's current pricing structure. While some cable-phone companies offer free local calls, the area is often quite small and calls to internet companies are often expressly excluded. To find out if cable is available in your street, call Cable Information on 0990 111777. The 'indirect access' firms such as ACC, Mercury and Dial 1602 do not have local networks and so you would always be using BT for your local calls even if you subscribe to any of them.

- As a result of BT's minimum charge of 5p, if all you want to do is send and retrieve your e-mail and you can do this in less than one minute, then it doesn't really matter what time of the day you use the internet.

- If you use BT, you should consider the potential savings by joining 'Friends and Family', (free to join; nominate up to ten numbers which can include the phone number for your internet connection; 10% discount on all direct dialled calls to those numbers, and 20% discount on one number you choose as a 'BestFriend') and PremierLine (annual membership fee of £24; additional 15% discount on direct dialled calls to UK and international numbers). If your quarterly telephone bill is more than £45, then it makes sense to use both schemes; this would mean that an hour on-line at the weekend would only cost in the region of 40p.

- People often assume that calls are cheaper on bank holidays but this is not always the case; for example, apart from certain days during Christmas and the new year, BT charge full rates for bank holidays.

- You do not need to spend a fortune on phone call charges; my busiest month on-line so far has been December 1997 when I was doing a lot of research for this series. I was also conducting an e-mail questionnaire, and had subscribed to four mailing lists and three newsgroups. I spent 7 hours on-line, (an average of 15 minutes/day), at a cost of £6.04 (about 80p per hour).

- Use the fact that the internet works by transporting small packets of information between computers. There are times when your modem connection will be waiting for some packets of information to arrive or it may be totally void of 'traffic', having finished downloading a page from the WWW. This spare bandwidth can be used by one of your other internet applications; for example you

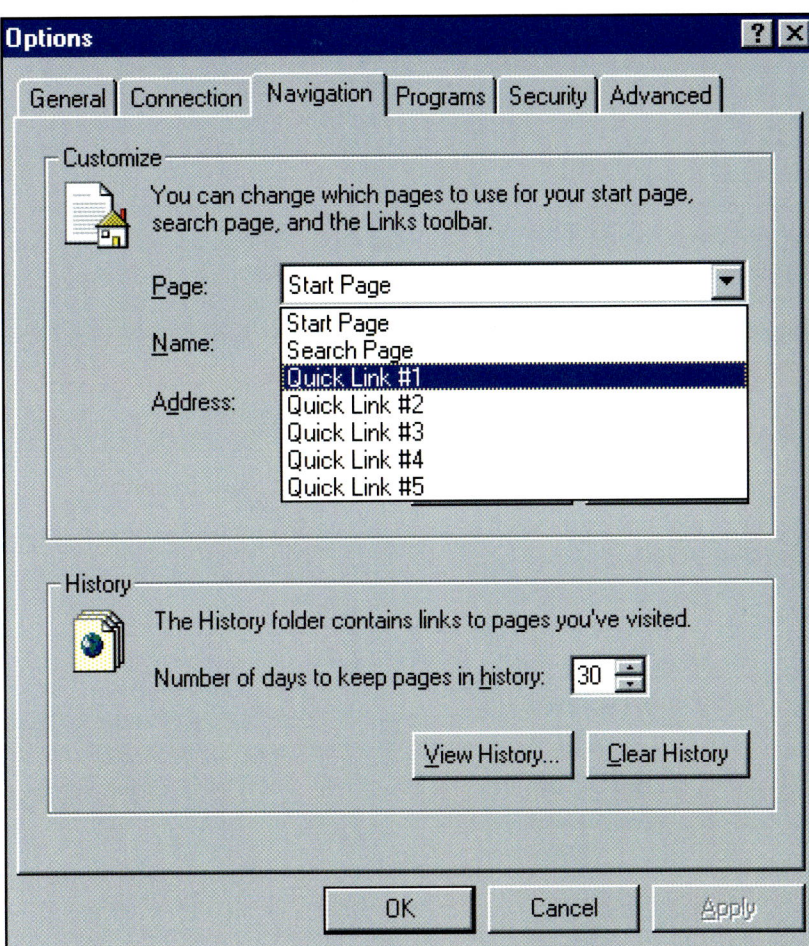

can be downloading your newsgroup articles and e-mail messages while browsing a web page; just remember that the data being transferred to your computer by the modem will have to be divided between the different tasks if there is no spare bandwidth.

- If a web page is taking a particularly long time to download, open a new window to carry on browsing while you are waiting: choose **File**, **New Window/Window** in MSIE or **File**, **New Web Browser/Navigator Window** in NN.

- Scan the subject line of incoming e-mail messages to see if there are any that would be better answered while you were still on-line. However, it is normally best to read and reply to the bulk of your e-mail messages and newsgroup articles off-line. Composing your replies off-line not only saves you money on telephone charges but also allows you more time to think about what you are trying to say. Things written in haste may later be regretted.

- What is a Cookie and should I accept it?
When you are browsing the WWW you will find that when you visit certain web pages, you will be asked if you will accept a cookie before you enter that page (you can switch off this alert in your browser). A cookie is a small piece of data that will, if accepted, be saved onto your hard disk. When you next visit that

Fig. 2 Screen shot showing the window in *Internet Explorer* version 3 where one can customise the Quick Links toolbar (Reprinted with the permission of Microsoft Corporation)

Bandwidth = the amount of data that can be transmitted over a connection in a fixed amount of time

FTP = file transfer protocol; a protocol by which data is either downloaded to your computer or uploaded to another computer on the internet

web site, your browser will send the cookie back to the web server.

Cookies are often used for various kinds of personalisation — the date you last visited a site or the particular web pages that interested you. From such data, the author of the web site can offer you personalised information when you next visit the site. Cookies are sometimes used to store your 'user name' and password when you register at a site.

Cookies can speed up your browsing and are pretty harmless, but if you feel that it is an invasion of your privacy, then simply do not accept them.
(The cookie is named after the fortune cookie — the sweet with a message inside.)
- Create a text file and name it something like *Useful text from the internet*. Whenever you come across some text on a web page that you want to keep (and you do not want to save the whole page), just copy and paste the text into the text file. This of course assumes that you have checked that you are not violating any copyright law. You can then read the text later at your leisure.
- Create another text file to record any 'user names' and passwords you devise as you register to join certain web sites and services. These can soon build up and if you only visit the site occasionally you will soon forget what they were. Many sites actually send you an e-mail with confirmation of your user-ID and password; keep these in a separate folder within your e-mail program.

Fig 3 The appearance of the FTP site when viewed through a web browser

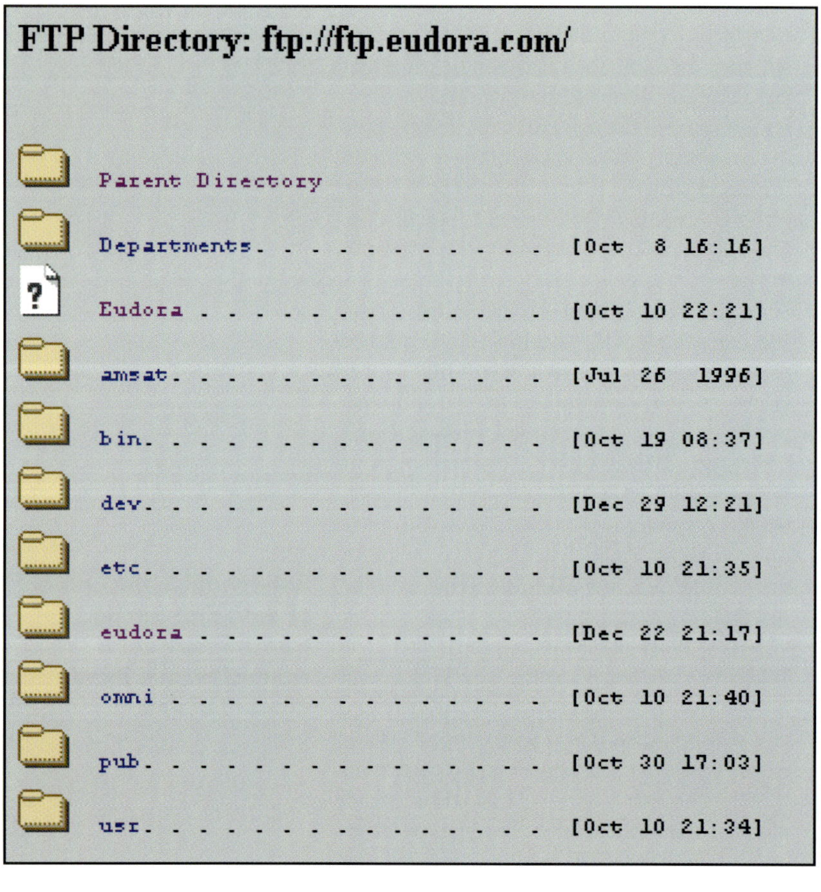

FTP Directory: ftp://ftp.eudora.com/

```
📁  Parent Directory
📁  Departments. . . . . . . . . . . . . - [Oct  8 16:16]
❓  Eudora . . . . . . . . . . . . . . . . [Oct 10 22:21]
📁  amsat. . . . . . . . . . . . . . . . - [Jul 26  1996]
📁  bin. . . . . . . . . . . . . . . . . [Oct 19 08:37]
📁  dev. . . . . . . . . . . . . . . . . [Dec 29 12:21]
📁  etc. . . . . . . . . . . . . . . . . [Oct 10 21:35]
📁  eudora . . . . . . . . . . . . . . . [Dec 22 21:17]
📁  omni . . . . . . . . . . . . . . . . [Oct 10 21:40]
📁  pub. . . . . . . . . . . . . . . . . [Oct 30 17:03]
📁  usr. . . . . . . . . . . . . . . . . [Oct 10 21:34]
```

- Create a directory on your hard disk where you can place any downloaded files from the internet. Just before you transfer the data, create an aptly named sub-directory in which you can store the files. Downloaded files often have obscure filenames and there is nothing worse than not being able to remember where you placed an important file. Carry out a virus check before running any files

Getting the best from your existing hardware and software
- Check your modem manufacturer's web site for upgrades to your modem's driver. Installing a new driver may improve your modem's performance.
- For tips about modems, there are many good sites worth visiting on the WWW.[6]
- If you are a BT customer, you can check for line noise by dialling 17070 and selecting 'Quiet line test'.
- If you have problems with obtaining a reliable high-speed internet connection, contact BT and ask them to check your telephone line. Sometimes it can just be a simple case of an engineer at the exchange increasing the signal-to-noise ratio for your line; this has the effect of reducing the amount of noise as a proportion of the total signal.
- BT's Select Service called Call Waiting does not work well with modems, since the warning bleep can cause the modem to disconnect. Switch off this service temporarily before connecting to the internet by using the code #43# and switch back on later with *43#. (It is also possible to turn off call waiting on your modem/computer settings.) The Call Minder scheme may be a better option — instead of interrupting you, this will intercept and record incoming calls at the exchange and then notify you once your internet connection is finished.
- When you use the WWW, your browser stores small chunks of your files across the surface of your hard disk. This clutter can eventually cause your hard disk to slow down as the information stored on it becomes more and more fragmented. If you are a frequent user of the WWW, routinely check to see if your hard disk requires defragmenting by using the Windows 95/98 *Disk Defragmenter* program found in **Start**, **Programs**, **Accessories**, **System Tools**.
- Although an FTP program is quicker than your web browser for downloading files, it can be quicker to use your web browser to view the contents of FTP sites. To view an FTP site, type the FTP address in the browser's URL address box, for example, typing **ftp.eudora.com** will show you the contents of the Eudora FTP site (fig. 3).
- Popular downloadable files are often stored at more than one site on the internet; these other

sites are called mirror sites. If you are given the option, select the site which is geographically closest to your computer since this will often produce the quickest transfer of data.

- Avoid using formatted e-mail; not everyone uses an e-mail program that can read it.
- If you need to send some information to a group of people and it contains large pieces of data (for example some tables and graphics), rather than trying to e-mail the data as attachments, place the information on your home web page and e-mail the appropriate URL to the people concerned. This can be easier, quicker and also gets around the problem of knowing what sort of attachment the recipient's e-mail program can handle, for example MIME, Uuencode or BinHex.
- Avoid sending e-mail to the wrong person. If you are the recipient of a message which has been sent to a list of people, make sure you send the reply to just the sender, rather than to everyone on the list; this may save yourself a lot of embarrassment. Check that just one name is visible in the **To**: box of your reply.

How do I avoid getting junk e-mail?

Unfortunately, junk e-mail, (unsolicited commercial e-mail), is becoming almost as bad as conventional junk mail and there is no easy way of avoiding it. The best advice is to simply delete it.

The problem can be greater if you post messages to newsgroups, since this is the most popular source of e-mail addresses for junk merchants. A popular way of getting around this is to falsify your e-mail address when posting to newsgroups, using something obviously wrong such as

paul@pdownes.demon.co.ukx or
paul@nojunk.pdownes. demon.co.uk

This should fool the automatic address-extracting software, while any person wishing to reply to the e-mail should notice the error in the address and be able to correct it. Put a note in your e-mail signature to bring this alteration to the attention of the recipient.

If your provider and e-mail program supports separate e-mail mail boxes, then another approach is to use a different name before the '@' for your postings to newsgroups (and also in your e-mail address in your browser). This will then help you distinguish responses to them from the rest of the e-mail you receive.

- If you use NN, not only can you speed up your browsing by turning off the auto-loading of images, (**Options**, **Auto Load Images** in NN3 and **Edit**, **Preferences**, **Advanced**, **Automatically load images** in NN4), but once you have found a page that interests you, just click the ◯ *Images* ◯

button and the page will reload with the graphics. The auto-loading will remain off when you visit the next page. Using MSIE, it is less easy to toggle between images showing and not showing.

What is a Cache?

Caching is a process of storing files you need often. Caching can help speed up the process of browsing because your browser, when seeking a particular web page, will first look on your caches to see if a copy of the page is already stored locally. There are two forms of cache:

Memory (RAM) cache: this only stays active during one browsing session. If you disconnect from the internet but leave your browser running, you will find that you can re-visit the pages that have been temporarily stored in your RAM.

(Hard) disk cache: this is a more permanent cache but cannot be used to browse off-line in versions 3 of NN or MSIE unless you use an extra piece of software (see later).

Both types of cache can be configured by your browser. In NN, check the size of the memory cache and disk cache by clicking on **Options**, **Network Preferences**, **Cache** tab for NN3 and **Edit**, **Preferences**, **Advanced**, **Cache** 'subdirectory' in NN4. If your computer has 8 Mb or more of RAM, the memory cache can be set to at least 1000 Kb. The size of the disk cache depends upon how much free space you have on your hard disk, but anywhere from 5 Mb to 20 Mb would not be unreasonable. In MSIE you can set the size of your hard disk cache to a percentage of your hard disk drive. In MSIE3 select **View**, **Options**, **Advanced** tab and click on the button labelled ◯ *Settings* ◯; in MSIE4, select **View**, **Internet Options**, **General** tab and click on the ◯ *Settings* ◯ button. You will need to clear out your disk cache on a regular basis; if it becomes full then your hard disk will slow down as it attempts to swap files to make space for new ones to be stored.

Problems with caching: the downside of caching is that you may be presented with an old web page from your disk cache rather than an up-to-date page from the WWW.

In both NN and MSIE you can configure how a page in your cache is compared with the actual page on the WWW; you will find the settings in the same place where you set the size of your cache. You can set your browser to always use the cached copy, check for a new version every time you request it, or else just once per on-line session. Using the cached copy is the fastest option. Getting it to check once per session is the best compromise for speed and currency of information. If however, you access news sites where the

World-wide web (WWW) = a subset of the internet, composed of millions of inter-linked pages, each of which can contain text, pictures, tables, sound or video

URL = uniform resource locator; the internet way of indicating the unique address of a particular page on the WWW

Newsgroup = a type of electronic discussions group

the internet guide for dentistry

Web browser = software program that allows you to view pages on the WWW

Hyperlink = the links that tie together the WWW. Clicking on a link within a web page could take you to another page or even run an element of multimedia such as a video

information is continuously changing, select 'every time'.

- Small monitor screens can make it difficult for you to view a whole web page at once. Experiment with different browser font sizes to see if you can obtain a good compromise between an easily readable font and getting more information on your screen.

- Turn off underlined links in your web browser; this will make a vast improvement in the appearance and readability of web pages.

- If you want to create a less cluttered browser, both NN and MSIE allow you to customise the layout of the various browser toolbars and even hide them all together. This will leave you with more of the web page visible on your screen at any one time. (In NN the settings for showing or hiding the toolbars can be found under the **Options** or **View** menu bar. In MSIE you show/hide the toolbar by the **View** menu bar.

What is a proxy?

Proxy servers are computers that sit between your computer and the web server whose pages you are accessing; they hold an archive of the most popular, recently accessed web pages. Proxy servers speed up the transfer of data since they only have to visit the web server if the requested page has been recently updated or is not already stored locally. Since most web servers are situated on the other side of the Atlantic, this can dramatically speed up the transfer of web pages.

Configuring your browser involves entering the name of the proxy server and the port number in the appropriate box (check with your internet provider for the appropriate details). To get to the proxy set up screen in NN3, select **Options**, **Network Preferences** and then the **Proxies** tab. For NN4, select **Edit**, **Preferences**, **Advanced** and click on the **Proxies** 'subdirectory'. For MSIE, select **View**, **Options/Internet Options** and then the tab labelled **Connection**.

MSIE and NN both support 'Automatic Proxy Configuration' which can be more reliable and faster than the standard configuration. Contact your internet provider to see if they offer this service, and if so, how to set it up.

More tips on efficient browsing

- Use the right click of the mouse to access many of the common commands in your web browser, for example **Add Bookmark** or **Add to Favorites**

- You can also save pages which you have not even visited: right-click the mouse on the link to the page you want to save and choose **Save Link As** (or **Save Target As** in MSIE).

- Instead of using the scroll bars to scroll through a large web page, use the Page Up and Page Down keys on your keyboard.

- When you are browsing through a large web page on a site which does not contain its own search engine, you can look for a particular piece of information by using NN3's ○ **Find** ○ button or select NN4's **Edit**, **Find in Page**. For MSIE, click on **Edit**, **Find** to highlight each occurrence of the relevant word on that page

- When you use your web browser to store 'favourite' or 'bookmarked' sites, do not always use the default text as the description for the site. For example, the default text may be something like 'Welcome to my Home Page', which will hardly help you to remember the site later on. Instead, edit the suggested text to read something like '20 tips for successful endodontics'.

- In both NN and MSIE, you can create separate folders in which to organise your favourite sites. NN's bookmarks are very easy to organise, but if you want a folder to appear at the top of the list in MSIE (which is ordered alphabetically), then add a '**!**', '**#**' or '**@**' in front of the folder name. Keep a spare folder as a place to store interesting sites; you can organise them into a suitable category later on, once you are off-line.

- If you type or copy a URL into your web browser but are told that the page does not exist it may be because you either typed in the address incorrectly, or the page on the web server has either been deleted or renamed. Try deleting parts of the URL address from the right-hand side until you get to the root address of the site; if you are still told that the page does not exist, then it is likely that the web site is no longer maintained or has been moved to a completely different address. For example, if you are unable to find a recommended site and have been told that the URL is **http://www.dentalschool.ac.uk/perio dept/quiz5**, delete '**quiz5**' from the URL and try again. If this still does not work, delete '**periodept/**' and hopefully this should take you to the home page of the dental school.

- It is possible to view a list of URLs that you have visited in the recent past; click to the right of the address bar to show the drop down list. (However, a URL will only appear in this list if you typed it into the address bar rather than just following a hyperlink).

- There is a shortcut method for typing in the URL of any web site that follows the common format **www.example.com**. Just type **example** into the address box and the browser will automatically fill in the rest of the address for you.

- Frames can often cause confusion when you are browsing. In NN4 you can get rid of a web site frame by right clicking within the frame and choosing **Open frame in new window**. In MSIE4 you can print an individual frame (rather than the whole page) by right-clicking on the frame and selecting **Print**.

- In MSIE4 you can retrace your steps more than just one page at a time. Click on the small downward pointing arrow next to the ○ **Back** ○ button and select any one of the previously viewed web pages from the drop down list (see figure 4).

- If you are using MSIE4 and you want to use Yahoo! to carry out a search, type **go**, **find** or **?** in the address bar followed by the word or phrase you want to search on. Yahoo! will automatically open and complete the search.

Using additional hardware and software

- If you are using a slow modem, consider upgrading to a faster one, (preferably 56 kbps). The cost is not extortionate and if you use the WWW a lot, you will soon recoup some of your outlay in reduced phone charges (unless of course you end up spending even more time browsing because it becomes a more pleasurable experience!). When it comes to browsing the WWW it is definitely true to say that 'minimum specification equals minimum performance'.

- If you find that you (and/or your family), spend a lot of time on-line, consider installing an additional telephone line. This will prevent your family, friends or emergency patients from complaining that they cannot contact you because your phone is always engaged. Prices at the time of going to press were £116.33 for installing an extra line plus the quarterly rental fee of £26.62; watch out for special offers on the cost of installation.

- Another option would be to convert your existing BT telephone line to BT's new Home Highway. The installed box has two analogue and two digital channels. You can use any combination of up to two of the four socket outlets at any one time. This means that you can either use the two analogue channels (eg telephone plus dedicated line for a modem or fax machine), one analogue and one digital channel (eg telephone plus 64 kbps digital connection) or use both digital channels from one socket to give you a 128kbps digital connection. The current cost for converting is £116.33 (inc. VAT) with a £40 monthly rental charge that has a £15 monthly call allowance. If you want to use the digital connection with your computer you also require an ISDN card (£50). At the moment, the major drawback for using the full 128kbps connection is that your internet provider is likely to charge you a higher monthly fee and you calls will cost twice as much because you are effectively using two lines at once.

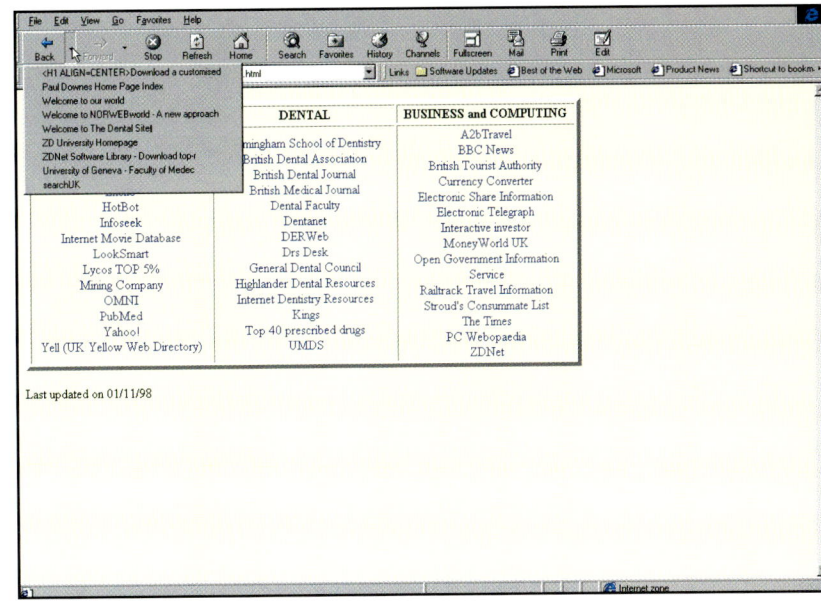

Fig. 4 This screen shot shows how MSIE4 allows you to retrace your steps and go back more than just one page by clicking the downward facing arrow, next to the ○ Back ○ button (Reprinted with the permission of the Microsoft Corporation)

- There are new internet providers springing up who are offering remarkable deals for connecting to the internet. The best value has to be Dixons' FreeServe; it costs nothing other than your telephone calls. However, there is no such thing as a free lunch; calls to their help desk are charged at £1 per minute and if the service becomes oversubscribed there is a risk that the quality of connection will suffer.

- The Microsoft IntelliMouse and the Logitech Pilot+ are mice that use an extra 'wheel' to scroll effortlessly up and down large web pages (and other Windows 95/98 compatible applications such as Word).

- To keep a track on exactly how long you spend on-line and how much it is costing you, download the *Phone Monitor* shareware program.[7] You enter the details of your telephone company charges (including discount schemes), and the program then produces a break-down of costs for any particular day or month. The current cost of registering the program is £5.

- Another useful internet utility program is *Bookmark Importer*.[2] This program is aimed at an audience who use both NN and the MSIE and wish to share bookmarks between the two applications. The current cost of the program is US$19.95.

- Remember that the internet is not the only way of obtaining cheap or free software. Many computer and internet magazines supply a cover-mounted CD-ROM that may contain most of the popular internet application software available; this can save you many hours spent downloading software from the internet. CD-ROMs are capable of storing 650 MB of programs; it would take days to download the same amount of data from the internet using a fast modem!

- For the same reason, do not download Section 63 CAL programs from the internet if you are

Freeware = software that is copyrighted by the author but made available to end users without charge

Shareware = software that you can try before you decide to buy

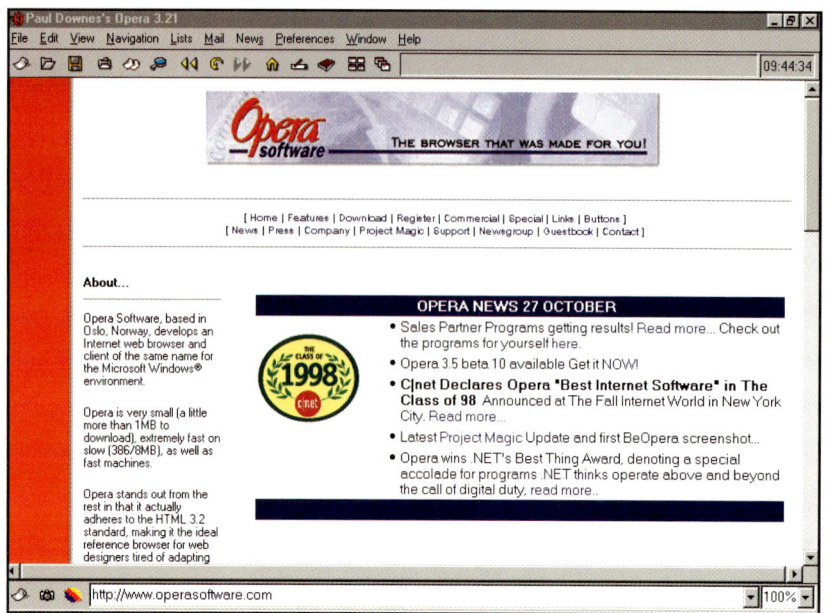

Fig. 5 The Opera browser competes against Netscape and Internet Explorer by being small, fast and robust

1. Survey on the use of the internet by UK dentists:
   http://www.pdownes.demon.co.uk/survey.html
2. *Bookmark Importer:*
   http://www.webobj.com/bookmark/
3. Customised start page
   http://www.pdownes.demon.co.uk/startpage.html
4. *WinZip:*
   http://www.winzip.com/
5. British Telecom:
   http://btshop.bt.com/
6. Introduction to modems:
   http://www-ccs.uchicago.edu/technotes/misc/Modems.html
7. *Phone Monitor* program:
   http://www.vizion.demon.co.uk/
8. Opera browser:
   http://opera.nta.no/
9. *WebWhacker:*
   http://www.bluesquirrel.com/products.html
10. *Teleport Pro:*
    http://www.tenmax.com
11. *Cache Explorer:*
    http://www.mwso.com/
12. *WebFerret:*
    http://www.ferretsoft.com/netferret/index.html
13. *InterMute:*
    http://intermute.com/
14. Companies producing anti-virus software include:
    Dr Solomon:
    http://www.drsolomon.com
    McAfee:
    http://www.nai.com/default_mcafee.asp
    Symantec:
    http://www.symantec.com
15. *Free fax* software:
    http://www.tpc.int/tpc_home.html

eligible to receive the programs free by post. The process of downloading, unzipping and installing the software can be quite a fiddle compared with loading it from a floppy disk or CD-ROM. Contact your local postgraduate dean for details.

• Avoid using 'bloatware'; over-large programs that contain many features that you are unlikely to ever use. Be selective when installing additional internet utilities; ask yourself whether it will be potentially useful and/or increase your productivity.

• If you own a low specification computer and you have problems running NN or MSIE, consider obtaining a copy of the Norwegian browser called Opera[8] (see figure 5). This robust and fast WWW browser is not so demanding of your computer's system resources. Indeed, many people now use Opera as an alternative to NN and MSIE; it automatically imports your existing bookmarks and favorites. The current cost of the program is US$35.00.

Off-line browsers are additional stand-alone programs which can be configured to visit a particular web site and download every page from the site to a pre-set number of levels, based on its interrogation of the links from that site. Once you are off-line you can then browse the entire site at your leisure. The downside is that they can soon fill your hard disk with a lot of information that you might not necessarily be interested in, and some-

times miss the piece of information you were looking for. Some sites also block users from using this sort of software. Examples of off-line browsers are *Webwhacker*[9] and *Teleport Pro.*[10]

Cache Explorer[11] this is a very useful shareware program that enables you to browse the contents of your disk cache while you are off-line. Versions are available for both NN and MSIE. As well as reducing the size of your telephone bill, it also allows you to view the contents of your cache with a Windows Explorer-style interface that makes the job of weeding out unwanted data very easy. The current cost for registering the program is US$18.

Both NN4 and MSIE4 feature a built in cache explorer.

• I use an excellent freeware program called *WebFerret*[12] that can send a search query to all the major popular search engines at once and then compile the results. The software only takes a few seconds to load, is very simple to use and is a fast way to find specific information on the internet.

• One shareware program I am using that really does speed up the process of browsing the WWW is *InterMute*.[13] It works by filtering out unwanted advertisements, animated images and more. It runs on Windows 95/98, works with any browser and costs US$19.95 for a registered copy.

• Having a virus-checking program[14] on your computer is a good idea whether or not you are connected to the internet. Viruses infect your system when you run a program or a macro (for example in Word) — they do not hide inside ordinary text files. There are a number of hoaxes circulating around the internet which claim that just reading an e-mail can give you a virus; this is NOT the case.

• Thanks to an international network of agreements between internet providers it is possible to send free faxes by the internet. Each local provider acts as a fax server for their own area code. For example, all faxes sent by the internet to the UK are sent to a Demon server that then performs the final telephone leg of the journey using its national local rate service. Demon foots the bill, and in return, advertises on your cover sheet. If you want to send more than just text, then you will require some additional software that can be downloaded free of charge from the WWW.[15]

10 Current and future developments

The use of computers, both at home and at work, is increasing all the time. In one UK survey it was estimated that nearly 40% of households owned a PC, (DTI/Spectrum; August 1997). The internet is a more recent phenomenon yet the number of adults in Britain with access to the internet has increased by 1 000 000 in the previous 6 months bringing the total to about 7 000 000.[1] It is estimated that around one in 25 of all households in Britain are now linked to the internet.

There are, however, a large section of our society who are very resistant to using new technologies. The Department of Trade and Industry conducted research into the varying levels of confidence toward new technologies and found that the UK population can be broadly divided into five groups of IT users and non-users (Table 1).

This survey is significant for dentistry, since many young professionals fall into the first category of 'enthusiastic toward new technology'. However, almost 30% of dentists registered in the UK are female, who according to the survey, would tend to fall into the 'unconvinced' category. The good news is that the number of British women as a proportion of the overall internet user base is continuing to increase,[1] representing around two in five of all current users.

Dentistry
New technological developments such as intra-oral cameras, digital radiography and interactive patient-education systems are changing the shape of dentistry. The internet may not be quite as visible to patients as the other examples but it is already having an impact on how dentists are trained and how

they will keep up-to-date with their continuing education in the future.

School leavers
Students enter dental school with varying degrees of computer literacy; at the extremes, some are computer aficionados while others are complete computer-phobics. Existing computer skills depend upon an individual's past experience of using computers at school and at home. As recently as 1996,[2] it was shown that, in the final 3 years of dental undergraduates, only 37% felt confident to carry out a written assignment using a computer. A recent survey, carried out in March 1998,[1] showed that 41% of children in Britain have a PC at home and 29% of children aged 7–16 had used the internet at least once.

The Government's National Grid for Learning[3] plans to network every school, college, university and library in the country by the year 2002 but it will be some years before all pupils leave school with some guarantee of computer literacy. It has been suggested that one way around this problem would be to make a certificate in basic computer skills[4] one of the requirements for entry into medical/dental school, thus freeing up valuable teaching time in the pre-clinical timetable.

Dental undergraduates
Because of the lack of computer skills in entrants to dental schools, there has been a move toward including a formal IT skills course in the dental curriculum. However, at the moment there is no agreed IT curriculum and the depth of each course varies widely from school to school. Some courses only cover topics such as word processing and using CAL, while others include training on using e-mail

In this section the current state of internet use in the UK is considered. A discussion is given on how dentistry, particularly dental education, is being influenced by the internet. There is also a look at present developments in internet connections, software and hardware.

Key:

Text that is of a more technical nature

web browser
Keyword defined in the margin

Table I The 'Information Society' of the UK according to research by the Department of Trade and Industry, August 1996

Group	%	Overall percentage taken from UK population Group description
Enthusiasts	25	IT users who are mostly young, male and well off
Acceptors	21	Users, but don't see the point in taking the technology home
Unconvinced	20	Don't see the point in it — high proportion of women
Concerned	18	Nervous about the technology
Alienated	16	Non-users, lower income and/or elderly

and WWW browsers. The more advanced IT courses tend to occur at those schools that are already involved in writing computer-based learning material.

In many dental schools, training is carried out by a small number of keen enthusiasts and few schools actually have a budget identified for teaching IT. In one recent survey,[5] the main obstacle identified by dental students as a barrier to using IT was the lack of adequate training. For these reasons it is likely to take some time before all UK dental graduates possess the necessary computer skills to equip themselves for a future of self-directed learning. It is well recognised that before any dental IT initiative can ever hope to be successful, it is essential that senior management in all the main dental institutions (dental hospital, community or postgraduate deanery) endorse and support basic computer training skills as well as the use of new technologies. Many will have to start by ensuring that they themselves acquire a certain level of computer literacy.

Dental schools, such as Newcastle, have started to harness the power of e-mail by creating a mailing list for each year of students (eg BDS98) and their teaching staff. It is used to clarify points raised after lectures, give additional information such as references, disseminate news flashes, and answer queries from students.

One innovative approach has been undertaken by King's Medical and Dental School who provide medical and dental students with their own laptop computers at discounted rates. This enables students to word process their course work and have access to the increasing list of electronic resources on the school's network. This could become more important now that Guy's and King's have merged, often resulting in students and teachers split between the two sites. Remote access of the network by modem could greatly help communication.

Many dental schools have placed their available CAL programs on the school computer network so that students can work through the various electronic modules at their own pace and go back at any time for purposes of revision. The student's progress can be stored electronically and be monitored by their tutors. Confidential information can be made secure by the use of passwords. On the MedWeb site[6] at the University of Birmingham, there is a Computer Assisted Assessment section that features databases of searchable multiple choice questions, some of which cover dental topics. The system will automatically mark your questions on-line and give you a score and feedback on your performance.

A good example of how the internet can be used as a management tool in undergraduate education can also be found on the MedWeb site;[6] the curriculum and timetable for the MBChB and BMedSC degree course are now

prepared and published electronically. The paper version of these documents are more than 100 pages long and are expensive to print and distribute, and even more expensive in time and resources to update. The electronic versions are much easier to maintain centrally and each individual teacher is responsible for their part of the curriculum. Each topic has links to other relevant electronic sources of information and the student can see where each topic will reappear later on in their course.

The WWW is seen by many teachers/lecturers as a great opportunity for increasing their teaching resources at very little cost. It is especially considered that the additional use of suitable multimedia elements, such as photographs and graphics can 'add value' to existing educational material. However, the sharing of resources has been slow to implement since many institutions are reluctant to share material without any guarantee of receiving anything in return. There are also potential problems with copyright, and arguments about the quality and appropriateness of other people's resources.

The UK's Academic Network is now connected to SuperJANET III; this is a high performance network that will provide the higher education and research community with a 155 000 Kbps connection that will greatly facilitate teaching and research.

General practice

Dental practitioners have been keen to embrace the use of new technologies to enhance patient care and this has included electronic communication. In the UK in 1997, 59% of dental practices used a computer at work,[7] and therefore a large proportion of dental practitioners already have some level of computer literacy. On October 14, 1998 there were 10 616 dentists with live contracts authorised to transmit EDI claims to the Dental Practice Board,[8] so there are already many dentists using a modem connection from work. What is not known is how many dentists in the UK are already using the internet and how this is changing. *The Dentists Register* is now published on the WWW,[9] and similar to the hard copy book, it contains just a postal address for each dentist; as yet there are no plans to include telephone/fax numbers or indeed e-mail addresses.

With the shift toward lifelong learning and an increase in the involvement in and responsibility for our own, self-directed education, the Department of Health has funded many CAL programs for dental practitioners.[10] These programs can be obtained on floppy disk or CD-ROM from your regional postgraduate dean, or downloaded from the WWW.[8] In the future, the internet could be an ideal way of distributing small CAL programs or updates to existing programs.

A new development has been the writing of CAL programs to run on a web browser. The WWW and browser technology offer an open platform for all computer users which means that people can access the information no matter what sort of computer they are using (eg IBM-compatible or Apple Macintosh). Dental CAL programs can be now be used directly over the WWW, for example the program 'Removal of Foreign Objects from Root Canals'[11] (fig. 1). Another innovation is the use of a CD-ROM to hold the main program plus large multimedia files (such as sound samples, photographs and video), while smaller, text-based files containing data such as updated statistics can live on a web site. Microsoft's Encarta encyclopaedia on CD-ROM is a non-dental example of how a CD-ROM can automatically link to pages on the WWW so that the very latest information is always available to the user.

For dental practitioners who missed out on being taught computer skills at dental school, there is the excellent Core Skills in Informatics in Dentistry (CoSID), that is available as part of the TEAMWORK initiative. This is a computer based course, suitable for all members of the dental team, that covers the basic skills needed to use computers, (including networks). There are also dental postgraduate courses being held up and down the country that deal specifically with computer skills training and using the internet.

The ease of 'publishing' information on the WWW has been somewhat detrimental to the overall quality of what is available. Dentists are used to reading refereed journals but we all need to develop an ability to think critically and question everything that we see on the internet. In the future it may be possible to configure browsers to access resources that satisfy trusted quality ratings such as Platform for Internet Content Selection[12] (PICS); think of this as a sort of kite-marking service.

The internet will never replace traditional formal postgraduate teaching; sometimes some of the best ideas from courses come from discussing issues with your peers over a cup of coffee during one of the intermissions. However, the way postgraduate courses are being run is already changing as a direct result of the internet.

A new project called PROVIDENT (Postgraduate Regional Online Videoconferencing in Dentistry) is currently running in the Thames regions. The project will assess the educational efficacy of using videoconferencing to supplement existing postgraduate education to dental practitioners. This involves linking two or more computers at remote locations by a fast ISDN digital telephone link and using a video camera at each location to send and receive live pictures and sound. Lecturers and clinicians from Guy's/King's, Eastman and The Royal London will be able to link with a

number of postgraduate centres and dental practices in the region. Although the system uses a direct call, it is possible to use an ISDN link through the internet provided one can establish a stable digital connection. This would greatly reduce the cost of calls if one wanted to extend the links to international speakers.

Some postgraduate deaneries, for example Wessex,[13] now place course details on their own web site. Eventually, it should be possible to produce a database of all postgraduate events in the UK and provide a search facility so that a dentist can find a relevant course by topic, speaker or venue, (for example, within so many miles of his or her postcode). The ability to contact the organiser or speaker by e-mail would allow the user to book onto a course and also make suggestions to the speaker as to what topic they would particularly like covered. It would also permit the user to contact the speaker after the event; the best question always seems to arise on the way home from the lecture!

Talks and conferences that deal specifically with dental computing now often provide a

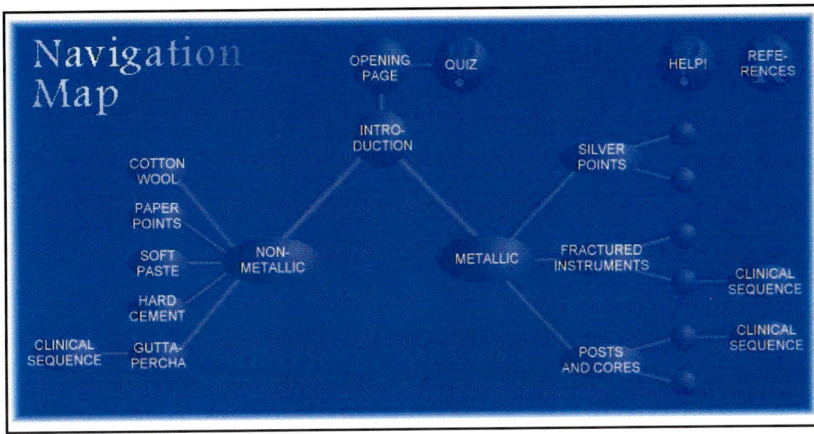

Fig. 1 The 'removal of foreign objects from root canals' CAL program. This shows the on-line navigation map

Fig. 2 The Schottlander web site runs a monthly on-line quiz using images from DERWeb Image Library

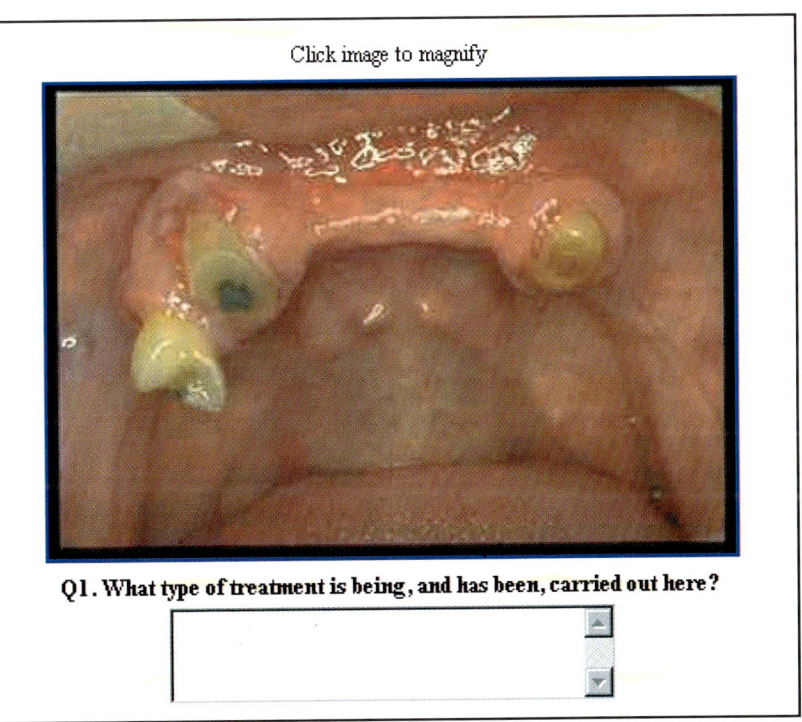

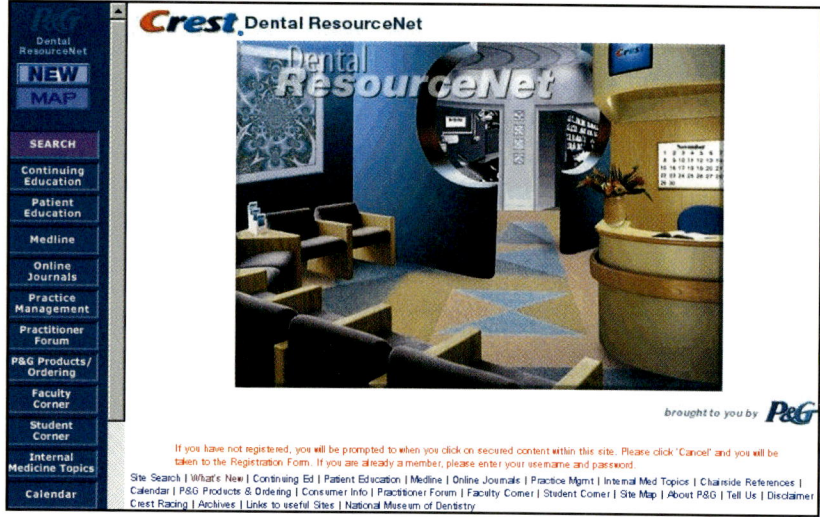

Fig. 3 The Proctor and Gamble web site showing links to its continuing education pages

'cybercafe' during the lunch interval for people to look at web-based resources and sites mentioned during the previous sessions.

Lecturers are increasingly using browser technology for their presentations rather than using conventional computer presentation software such as Microsoft's *Powerpoint.* The advantage of using a browser is that it is easy to construct a non-linear talk. Simple web pages are fairly straight forward to create and most modern word processors allow you to save a file of text as a web page. The different web pages can then be stored on the hard disk of the computer and be read by a web browser. The cost of hardware for either scanning existing photographs or cameras for taking still digital images have dropped dramatically in the past year and this makes it easy to add photographic images to presentations.

Next year, the British Society of CAL in Dentistry are planning to hold their annual conference as a virtual conference that will include digital presentations delivered by the WWW in the format of slides, text, sound and hyperlinks. A question and answer session, using the same

WebBoard software as is currently used on DERWeb, will follow each presentation.

Commercial dental companies have been quick to produce web sites that include an element of postgraduate education. Examples include the Schottlander Quiz[14] that offers the opportunity to win £100 worth of Schottlander products each month (fig. 2). The Proctor and Gamble site[15] offers formal continuing education credits for North American dentists, and offers short courses on such topics as 'The avulsed tooth' and 'The immunological and inflammatory aspects of periodontal disease' (fig. 3). With the increasing likelihood of compulsory reaccreditation and recertification for dentists in the UK, it has been proposed that up to 35 of the 50 hours of CPE should be informally structured and cover as wide a spectrum of educational activities as possible, presumably including educational activities through the internet.

Another idea may be to have structured on-line study sessions where groups of dentists could download a specific article, read it and later meet back on-line to discuss it.

One section 63-funded CAL program on oral ulcers has just been released on CD-ROM[16] and is accompanied by an on-line web site that allows the user to take a final competency test. To access the on-line test you require a password that you acquire by completing questions from each section on the CD-ROM. Successful completion of the on-line test is recognised by the Royal Colleges of Surgeons and the Faculty of General Dental Practitioners and will earn the person continuing education credits.

It is difficult to predict to what extent dental practitioners will ever use an internet connection during work hours. As well as the issue of cost, it is hard to imagine many dentists finding time to sit in front of a PC screen in addition to carrying out clinical work on their patients. However, there are three pilot schemes currently running that show that this avenue is already being explored:

- The TeleDent project[17] is undergoing a 1-year pilot study into the use of videoconferencing in postgraduate dental education and orthodontic diagnosis. Professor Chris Stephens from the University of Bristol is collaborating with a number of practitioners based at a dental practice in Bristol to deliver orthodontic consultations and treatment advice with and without the patient present.
- The new multimedia development facility at King's College uses an ISDN link to a dental practice on the Isle of White to enable large numbers of students to observe practice management and communication skills within a real live practice setting. There are potentially huge savings over such things as the time spent travelling and travel expenses.
- Doctor's Desk[18] is a pilot scheme for medical

Fig. 4 The 'Doctors Desk' web site has links to guidelines and other useful sources of electronic information

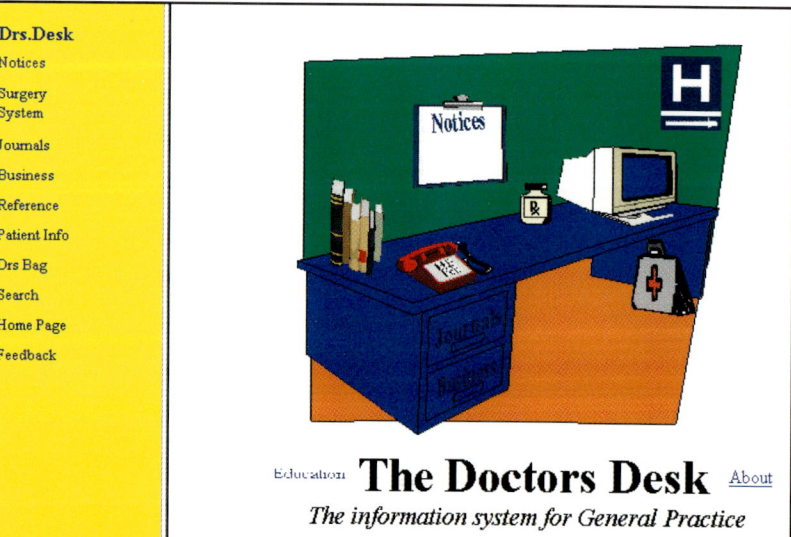

practitioners (fig. 4). It is as an example of a user-friendly front end to the internet and aims to give the doctor a quick and easy access to useful electronic sources of information and guidelines. The pilot evaluation has shown that most doctors use it during the working week, from 11–1pm and 3–6.30pm. This would suggest that it is possible to see patients and still find time to use the internet, although like the other two projects mentioned, it also relies on a fast ISDN connection. The system is linked to the medical practice clinical computer system as well as the doctor's local hospital computer system and uses the NHSNet to ensure that any patient data transmitted is secure.

A number of on-line versions of existing dental journals are available on the WWW; these are greatly enhanced by providing links to other related material. For example, the *British Medical Journal* now publishes the full text and graphics from the paper journal on-line. Most of the article references are active hyperlinks that either take you directly to the article or to a MEDLINE abstract. This negates the time-consuming and expensive task of tracking down and photocopying references. Click on an author's name and another hyperlink provides you with a MEDLINE list of other papers published by the author. There are hyperlinks to related articles and even an option that enables you to be alerted by e-mail when a new article cites the article you are reading. The other advantage of electronic publishing is that it is possible to include any follow up correspondence directly with the original paper. These examples illustrate the real power of electronic publishing and the weaving of links between like resources. Non-members of the *BMJ* have free access to the web site until at least the end of the year. If you have not yet visited this site then I recommend that you do so soon.

By the year 2000, our medical colleagues will have their practices connected to the internet through the NHSNet. In the future this may prevent the embarrassing situation in 1995 when women inundated their doctor's surgeries for advice about the link between the contraceptive pill and cervical cancer. The story had been front page news in every daily newspaper that morning but it took several days for every doctor to be contacted by post with recommendations on how to advise their worried patients. As a consequence of women not being able to get immediate advice, many stopped taking the pill and this resulted in a large increase in unwanted pregnancies.

There have been many occasions in the past two years where it would have been advantageous for dental practitioners to have immediate access to new regulations, guidelines or recommendations. A recent example is the General Dental Council's changes to its ethical guidance in respect of general anaesthesia. The changes were approved at a meeting on Tuesday, 10 October but the letter of formal notification did not arrive on my doorstep until Saturday 14 October. In the meantime, the media reported on the changes (incorrectly, I might add) on the Wednesday. This would have left 3 days where, had it not been for the UK GDP mailing list and the GDC web site, I would not have been able to give an accurate answer to my patient's questions about this matter. The time must surely be close when a central database of dentist's e-mail addresses is kept for the purpose of urgent communication.

Developments in internet connections

One of the things that is currently limiting the full use of multimedia on the WWW is the lack of bandwidth (the amount of information or data than can be sent over a connection in a given period of time). At the same time there has been a tremendous growth in traffic over the WWW because of the increasing number of users. In moments of frustration the WWW is sometimes jokingly referred to as the 'worldwide wait'.

At the present time most home users are restricted to accessing the internet by modem over an ordinary telephone line. Even the fastest modems (56 Kbps) actually only run at about 46 Kbps because of limiting factors inherent in noisy telephone lines. ISDN provides 128 Kbps and is now more affordable and flexible since the introduction of British Telecom's Home Highway. There are other mechanisms of connection that may provide future users of the internet with more than adequate bandwidth, (pessimists will argue that the content and number of users will simply grow to fill the available bandwidth):

- Asymmetric Digital Subscriber Line (ADSL)[20] technology can use existing copper telephone lines. It is designed to maximise the speed of downstream transmission since the majority of internet traffic is one directional (ie data in the form of web pages are transmitted from the WWW to the home). It may provide an 8000 Kbps connection but will require a special device to be fitted to the user's telephone system. Another version, called Consumer DSL (CDSL), does not require the extra telephone hardware and could give a maximum connection speed of 15000 Kbps. Although still in the development phase, these two options will probably be the cheapest short-term solution to speeding up data transmission on the internet.
- Cable internet operates at a theoretical maximum of 10 000 Kbps maximum, but since the bandwidth is shared between the number of customers who are using the system at any one time, it is more likely to give the end user a connection rate of about 600 Kbps. It could

Hyperlink = the links that tie together the WWW. Clicking on a link within a web page could take you to another page or even run an element of multimedia such as a video

Bandwidth = the amount of data that can be transmitted over a connection in a fixed amount of time; measured in Kbps (thousand bits per second)

take 30 to 40 years to replace the world's existing copper network of wires with more modern fibre cables.

- NORWEB/Nortel[21] have developed a new technology called Digital PowerLine that allows data from the internet to be transferred over normal electricity lines at speeds of over 1000 Kbps. The first public installation took place in December 1997 at a primary school in Manchester.
- Project Oxygen[22] is an ambitious new super internet involving a single 275 000 km fibre-optic cable with 38 distinct loops around the globe. It promises to deliver data transfer speeds at a mind-boggling 640 000 Kbps (ie 5000 times more bandwidth than existing ISDN). Provided the £8.7bn to fund the project can be raised, the first part of the connection should be up and running by the year 2003.

Developments in internet software

The internet has revolutionised how software is now developed, tested, marketed, sold, distributed and upgraded.

A beta-version of an application is one that is made available prior to the official release for the purposes of testing. It is now common practice for Microsoft to release beta-versions of some of their software onto the internet and ask the public to act as unpaid guinea pigs to check the software for faults.

The huge fight for supremacy in the WWW browser market has led to both Microsoft and Netscape giving away their browsers free of charge; a major company giving away its software was unheard of 5 years ago. Microsoft has come under a lot of criticism for combining the Microsoft Internet Explorer browser with the Windows operating system. The US Department of Justice and the European Union have been examining claims of anti-competitive behaviour and predatory pricing; at the time of writing the case has only just started being heard in court. In the relatively short time that Microsoft have become involved with the internet, they have had a huge influence on how it has evolved and this is likely to continue for some time.

On-line shopping has been slower to take off than first expected but one of the most popular items sold is, not surprisingly, computer software. In addition to the thousands of shareware and freeware programs available, a lot of the big commercial software firms now use the internet to sell and distribute their products.

One very useful innovation is the ability to obtain upgrades to commercial software through the internet, for example the anti-virus programs rely heavily on this to ensure that users can obtain maximum protection against newly discovered viruses. It is also possible to buy programs that will automatically find and download upgrades and bug fixes to much of your existing software.[23]

Advertising is now playing a more prominent role in the development of internet software and one example is *LifestyleFinder*.[24] This web application provides on-line businesses with useful consumer information, enabling them to target their marketing. Unlike many similar products, the user's privacy is not compromised, because they do not need to reveal specific information about themselves, such as gender, income or postal code. When a user logs onto *LifestyleFinder*, they are asked a series of lifestyle-probing questions. Based on the answers given, the program then suggests a list of WWW sites featuring products, services and places that are likely to match the user's preferences.

Current internet software buzzwords explained:

- *ActiveX:* a loosely defined set of technologies developed by Microsoft that enable a web browser (specifically Microsoft Internet Explorer), to play special multimedia files directly within the browser.
- *Java:* a programming language developed by Sun Microsystems. What makes Java special is that it can run on almost any type of computer, making it particularly well suited for use on the WWW. Small Java applications are called Java applets and can be downloaded from a web page and run on the user's computer by a Java-compatible Web browser, such as Netscape Navigator or Microsoft Internet Explorer. A Java applet can be an animation, a movie clip that plays automatically or a form that gives you immediate feedback.
- *JavaScript:* a simple coding system that allows web page creators to embed Java applets into web pages.
- *Plug-ins:* third party add-on software that adds new features to a commercial application. For example, there are a number of plug-ins for the Netscape Navigator browser that enable it to display different types of audio or video messages.
- *Push technology:* The ability to send data to a user without the user requesting it. The WWW is based on a pull technology where the browser must request a web page before it is sent. Broadcast media, on the other hand, are push technologies because they send information out regardless of whether anyone is tuned in. Increasingly, companies are using the internet to deliver information push-style. One example is PointCast,[25] that delivers customised news to users' desktops.
- *Streaming:* the ability to start playing the first part of a file while the rest is still downloading eg Shockwave's[26] sound and animation, Progressive Network's[27] RealAudio and RealVideo.

1   NOP research group's internet user profile study:
http://www.nopres.co.uk/

2   Chadwick R G. Basic IT skills of dental undergraduates: a case for supplementary tuition at university? *Med Teacher* 197; **19**:148-169

3   National grid for learning:
http://www.uknetyear.org/

4   RSA Examination Board. *Computer literacy and information technology.* 2nd ed. Heineman.

5   Grigg P A, Stephens C D, Davis N. *A survey of the IT skills and attitudes in final year dental students at Bristol University in 1996 and 1997.* British Society for CAL in Dentistry Conference, 1998.

6   Medweb:
http://www.medweb.bham.ac.uk/teaching.html

7   *Dental computer survey, 1997.* Dental Practice Board, Eastbourne, E.Sussex.

8   Dentanet:
http://www.dentanet.org.uk/

9   General Dental Council Dentists Register:
http://www.gdc-uk.org/

10 BSCD list of Section 63 funded CAL programs:
http://www.derweb.ac.uk/s63cal.html

11 Removal of foreign objects from root canals CAL program:
http://www.bham.ac.uk/dentistry/hot/

12 Platform for internet content selection:
http://www.w3.org/pub/WWW/PICS

13 Wessex postgraduate courses:
http://www.wessex.org.uk/dental//index.htm

Developments in internet hardware

Internet TV boxes are now available in the UK and allow access to the internet using a standard television and telephone line. The NetStation costs £300 and it uses NetChannel as the internet service provider, at a charge of £14.95 a month. At the moment, 'Internet TV' gives only limited access to all the features of the internet compared with a computer but this may improve with further developments.

This autumn, the UK will be the first country in the world to launch digital terrestrial TV and all that will be required to access it will be a set-top box costing around £200, (this converts the compressed digital signal into television reception). Digital television will allow interactive channels for shopping, home banking and games as well as an internet service.

Household appliances may eventually become connected to the internet; for example, Spyglass[28] are developing software that would eventually make it possible for a web page to remotely program your video recorder or to turn on your central heating.

Gazing into the crystal ball

There are very few certainties in this world but one of them must surely be that some parts of the internet will change almost beyond recognition during the next five years.

This is because the whole system is still evolving and new uses for this technology are being discovered all the time. The internet is developing because people are trying things out; there is still no overall plan.

We live in very exciting times with the internet having an enormous impact on our daily life, both at home and at work. Now is the time for all members of our profession (young and old, male and female, clinical and non-clinical) to fully embrace the internet and what it has to offer. Wake up and smell the coffee!

The author would like to thank the many people who have given helpful feedback to the draft manuscripts, especially David Speechley who scrutinised the whole series. The author is also grateful to all the GDPs who returned their e-mail questionnaire; the results were very useful in structuring the series.

14 Schottlander quiz:
http://www.schottlander.co.uk/
15 Proctor and Gamble:
http://www.dentalcare.com/
16 Oral ulcers CAL program:
http://www.eastman.ucl.ac.uk/~ulcers/cd.html
17 Cook J. *ISDN videoconferencing in postgraduate dental education and orthodontic diagnosis.* Proceedings of the CTI medicine learning technology in medical education conference, 1997.
18 Doctor's Desk:
http://www.drsdesk.sghms.ac.uk/
19 *British Medical Journal:*
http://www.bmj.com/
20 ADSL Forum:
http://www.adsl.com/
21 NORWeb:
http://www.norweb.co.uk/index.htm
22 Project Oxygen:
http://www.oxygen.org
23 Oil Change:
http://www.cybermedia.com/products/oilchange/ochome.html
24 LifestyleFinder:
http://www.lifestyle.cstar.ac.com/lifestyle/
25 Pointcast:
http://pioneer.pointcast.com/
26 Shockwave:
http://www.macromedia.com/shockwave/
27 Progressive Network:
http://www.prognet.com/
28 Spyglass:
http://www.spyglass.com/

Index